The Quantum Kinetic Chronicles

Cover Images by McKane Lee

Printed in the United States of America.

The Quantum Kinetic Chronicles

The Discovery of Cold Nuclear Fusion

Episode I
The Quantum Kinetic Well

Episode II
Journey Beyond the Quantum Horizon

McKane B. Lee

Riley D. Lee

CONTENTS

Preface

In the chronicle of scientific history, departures from established doctrines and customary areas of research have at important junctures inaugurated transformational concepts that advance human progress.

In the present narrative, the trajectory of human development has been impacted by the disruptive discoveries of a solitary inventor and outsider, a young man shunned for a decade by scientific authorities, academia, and industry leaders. Despite revelatory experiments, exquisite expository writings, published research, historic press releases, and inexhaustible efforts to reach out to government agencies and private industries, he found himself wandering — even exiled at times — in a scientific wilderness. Nevertheless, he persisted with a singular purpose — not for personal fame and fortune — but to pursue discoveries that could ignite a new age of technology.

This book is a true account of a young American inventor on an intrepid journey off the beaten path of accepted science. A man who, against all the odds, successfully harnessed the power of transmutation at the level of the electron, the most fundamental constituent of matter. It describes how his unconventional work with editing nuclear elements brought ancient alchemy out of the realm of mythology and into the material world of scientific measures where the technology can now be used to heal a planet.

The Quantum Kinetic Chronicles include Episode I, *The Quantum Kinetic Well*, and Episode II, *Journey Beyond the Quantum Horizon.* Taken together, the *Chronicles* are evocative of modern lore revealing the humble and unconventional origins of Apple, HP, Amazon, Disney, Google, and Microsoft. As such, the *Chronicles* provide more evidence the world's greatest problems can be solved in someone's garage.

Episode I is a probing analysis of the ingenuity of a next generation inventor who — after a near fatal fall from a mountain — locked himself in a garage for six years with the implausible goal of stopping climate catastrophe. Following the path of sacrifice, he forfeited a marriage, his coaching career, and a restorative relationship in Hawaii, to zero in on untouched and unrealized truths about the universe.

His persistent and thorough research in a neighbor's garage led to a paradigm-shifting discovery and invention of the world's first cold fusion nuclear reactor. The monumental machine provides a framework for a new clean energy system based on harvesting vast amounts of energy from any water source.

Episode II continues the saga of sacrifice as the young American innovator aimed at unveiling his disruptive technology to the world. Opposed and shuttered off by governmental authorities, academia, and industry leaders, the lone inventor persisted on his riveting quest by taking his fusion reactor inside the world's most renowned research laboratories where bold experiments led to confirmation of an unprecedented technology.

The *Quantum Kinetic Chronicles* are the authors' offer of proof that the future of humanity will be driven by transformational concepts rather than by big money powerbrokers and legacy institutions seeking to maintain their establishment niches.

Dedication

This book is dedicated to David Butler, true friend of the inventor

David was born on the same day of the year as the inventor

David died on the anniversary date of the inventor's first discovery

About the Authors

Formerly a teacher and digital media professional with Electronic Arts, McKane Lee graduated with a Masters in Digital Media from Simon Fraser, Emily Carr, BCIT, and UBC. The Quantum Kinetics' self-taught inventor and missile-man leads with his imagination. Lee is always thinking outside the traditional physics box. He pushes the envelope of what's possible in anything he puts his mind to. Lee holds the All-Time Highest Record in the Pole Vault for a Sophomore in Washington State (15'9'') and cleared 16'8" as a high school junior. He later entered the University of Washington as a D-1 scholarship athlete. In his third year at UW, he was rated a top five vaulter nationally before injury permanently sidelined his athletic career. Today his hard work and imagination in pushing boundaries have resulted in a new scientific paradigm — sustainable Cold Nuclear Fusion, the *ARC REACTOR* technology. Pablo Picasso was right about the power of unconventional thinking, "*Learn the rules like a professional, so you can break them like an artist.*"

Before Riley Lee became a writer/biographer, he was an Attorney/CPA with over 33 years of experience in civil litigation, business, and tax law. As a trial lawyer, he won several top jury verdicts and settlements in Washington state. Prior to his legal career, Mr. Lee was a Clinical Psychologist for 20 years and also worked in human resource administration including as a county-wide Mental Health Manager. In recent years, Mr. Lee has transitioned from careers in psychology and law to become an author, utlilizing his diverse professional background, understanding of human psychology, and legal writing skills to delve into the dynamics of scientific discovery at the level of the individual inventor.

1

A Fateful Road Trip

McKane remembers Happy. She regularly scolded him for spending too much time on his experimental, hydrogen-powered car out in the garage.

It's not going to work.

Your BMW engine is just fine the way it is.

It gets good mileage now.

What are you running from?

Can you come out of the garage for a minute?

He recalls one of his dad's quips about the two secrets to a long marriage. One is learning to say, "Yes dear," and the other is making enough money to keep her in the style to which she is accustomed.

But McKane is a firefighter. He isn't planning on making a fortune any time soon. So, he'll have to content himself with working on the first secret.

"Yes, I'll be right in, honey."

He and Happy were friends in high school but never officially dated. Happy grew up in a communal setting on a 291-acre ranch shared by 350 members of the Love Israel Family. It was a hippie community born out of the counter-culture movement in the 1960s. Happy's father, a clinical psychologist, had been part of the academic arm of the movement in California. His associates were famed psychotherapists and authors, Carl Rogers, and Fritz Perls.

Happy grew up accepting that she had 100 brothers and sisters on the ranch. Her bedroom was a small, unfinished cabin with a dirt floor. Every summer, the cabin was invaded by ladybugs, the source of her uneasiness around insects. She got involved in gourmet cooking early. The food supplied small eateries run by the family in Snohomish County. She also fashioned wonderful crafts to be sold at the annual Garlic Festival, attended by thousands.

He recalls having lunch with Happy in the high school cafeteria just before their graduation. Present as usual were his coterie of female admirers sitting too closely together at the round table. Happy manages to slip in next to him. She gets his attention long enough to whisper in his ear, "Let's meet again back here when we are both 30 and get married." He agrees to the proposed rendezvous, thinking it impossible to be single in twelve years.

They have been married for two years now. They live in a small rental house on a road with an Indian name: Sioux Drive. Family lore has McKane inheriting Indian blood on his mother's side.

He works at the fire station in Stanwood. She works in Bellingham as a food scientist for a large nutrition company.

They are getting along and still in love. He is thinking it is too soon for children.

* * *

Happy is planning a weekend road trip to Leavenworth. McKane will come along, even though he doesn't like driving in the snow. She hopes Alex and Kaitlin will come, too. Alex works with McKane at the fire station. Kaitlin works in nutritional health at Skagit Hospital. The two couples have everything in common.

Happy plans for the group to celebrate Alex's birthday on their brief mountain holiday. She is thinking they could arrive in Leavenworth after work on Friday night and catch a hotel room. The next day would be open for tourist activities and for viewing the Christmas lights, still up after the holiday season. She can drive her Subaru. It has four-wheel drive.

No, she doesn't need snow tires for the Subaru.

The sun is shining on a frosty Saturday morning in Leavenworth. McKane, Happy, Alex and Kaitlin are following a snow-covered sidewalk from the hotel into town. The respective couples walk, arm in arm, in the frigid air. They notice gift shops, craft stores and small eateries along the way.

They stop at an outdoor beer garden and order the obligatory stein of beer. Next, they pass by *A Matter of Taste*, a Bavarian gift shop and cafe. McKane and Alex walk in. They notice a taunting sign of warning on the front counter next to a hot sauce bowl: *Hotter Than Hell. Try at Your Own Risk.*

They accept the challenge. McKane is first to take the plunge into the hot sauce with pretzel in hand. Alex, not to be outdone, follows his buddy. A moment later they are dashing across the street to the Gelato shop for oral first aid.

In the evening the couples find an Italian restaurant. They order too much pasta to finish. In the waning sunlight they linger at the table talking about the day. The evening continues with a walk along a boulevard framed by festive lights of the season. They stop in at an old-fashioned, Bavarian drinking establishment where they enjoy sample-sized glasses of beer from local breweries. Before long, the platter is full of empty glasses.

A young man approaches. He gives Happy a big bear hug, then turns to introduce himself to the others. “Hello, I’m Sam, an old friend of Happy’s family. ” Sam lives nearby, not far from Leavenworth, and at a higher elevation. Sam officially invites them to Sunday morning brunch on the mountain.

Happy accepts without consulting the others.

* * *

McKane wakes early the next morning. He hadn’t slept well. He wonders if it was the alcohol or a frightening dream. Maybe it’s both. He presses to recall the dream, sensing it may contain an embedded message, perhaps an insight or even a warning. His search only reveals darkness.

Darkness, swallowed by a deep abyss.

Unsettled by the disturbing vision, he tightly closes his eyes and draws the covers over his head wishing to escape the looming black hole.

The others are still sleeping.

He thinks about their original itinerary for the day: check out at 11 a.m. and then a two-hour drive back home through the mountain pass. Happy’s phone lights up—a text. It is Sam confirming the invitation. She replies, “Yes.” McKane is startled by the unilateral change to the group’s initial plan.

It is growing colder in Leavenworth. Outside, light snowflakes are beginning to fall.

McKane checks the weather forecast for Stevens Pass. He opens the weather app on his cell phone. The forecast calls for heavy snow and winter storm warnings. He rolls over to tell Happy and sends out a text message to Alex and Kaitlin. He argues that with the approaching weather system it is safer to head west and home rather than east to higher altitudes. Happy insists that she is an experienced driver in the snow. She adds that her friends are expecting them. Not wanting to disappoint Happy or Sam, the others agree.

An hour later, Happy pauses the Subaru at the base of the ridge road leading to the ranch house.

It is an old logging road, converted to community access. It looks rough and steep. There are sharp switchbacks every few hundred feet. The air is even colder here than in Leavenworth. The wind is stiffer. The snow is still falling.

Heavy snow.

McKane says they should just turn back. Alex and Kaitlin are looking at him nervously. Happy is concerned, too. She texts messages Sam. He answers, offering to pick them up in his retro-fitted, four-wheel-drive truck equipped with chains, sandbags, snow tires, and a snowplow.

But she doesn't convey Sam's offer to the others.

Happy's Subaru ascends the ridge road through swirling snow. As it moves upward, the SUV encounters an ever-steepening slope. The slate gray skies are darkening by the moment. McKane notices accumulations of snow building on the road. The Subaru is doing well even without snow tires or chains. They will probably make it. The lodge is near the top of the ridge.

They arrive unscathed.

Greetings and introductions buzz with energy. McKane feels distracted. When it is possible, he escapes the happy chatter and meanders into the lodge's great room. He finds himself facing windows that reveal panoramic views of the valley below. He watches distinctly layered cloud formations approaching over the adjacent peak. The wind pulsates the windows. The front is moving in fast.

Soon, a blizzard encircles the house.

Given their uncertain arrival time, preparations for brunch haven't begun. And of course, they had to catch up on current events first. The well-planned gourmet brunch will be late. The chef of the outfit, Happy takes charge of making a delicious quiche.

McKane notices again the drifting snow.

After brunch, McKane contrives excuses for announcing their early departure.

Meeting you has been wonderful.

We would love to stay longer, but maybe next time.

We all work tomorrow.

We really need to get on the road.

* * *

The descent down the ridge road seems to take forever. McKane and Alex are positioned in the back seats. Kaitlin is the front seat passenger. Happy drives cautiously. After all, she is winding through an 8,000-foot mountain range after a heavy snowfall in a small SUV without chains or snow tires.

As the downslope steepens, Happy manages to keep the car's momentum under control by gently pumping the brakes. She is even more deliberate at the sharp switchbacks. The Subaru is moving in slow motion when it encounters an impossible turn on the icy corridor.

Suddenly, the SUV is flying, catapulted into a white abyss.

The car careens hundreds of feet down the steep landscape. Kaitlin begins screaming when the car severs the first tree, then another.

The car somersaults.

Glass is shattering.

Metal is creaking, loudly.

The engine is racing out of control. Untethered objects are airborne within the chamber.

McKane remembers the SUV's roof getting smaller and smaller with each roll-over. They are trapped.

Is this how it is going to end?

Maybe this is a nightmare.

He tries to wake up, but his effort only confirms the terrible truth.

This is real. Our lives will soon be over.

He pushes against the collapsing roof with all his strength. The Subaru is still shrinking. He counts the roll-overs until it gets to twelve. *On the thirteenth, we will all be dead.* The car suddenly stops upside down against a tree, quiet and motionless in the cold mountain air.

The resting spot...

Somehow, he is alive. So are the others.

* * *

McKane exits the car, blood streaming down his forehead. His face is throbbing, beaten hard by loose, flying objects in the car. Happy's jars of canned vegetables, homemade jams, and jellies. His back, shoulder, and neck send agonizing messages of concern.

Despite the injuries incurred, his training kicks in, directing decisive action in the moment of danger. His full focus is rescue.

McKane makes his way to the others still trapped within the mangled wreckage. Quickly, he checks medical status to see if they can be safely moved out of the vehicle. First, he evacuates Happy and Alex from the devastation. Kaitlin cannot stand up. Her leg is badly injured.

Alex and McKane work together to pull her out of the vehicle.

Then McKane calls 911.

* * *

The local firefighters arrive on the scene. McKane asks if they are ready to perform a "high angle rescue."

They cannot comprehend the question. "What is a high angle rescue?" they ask.

"We are only volunteers," they explain apologetically.

Stunned by the admission, McKane realizes there is no option but to take command of the situation. He directs the volunteer firefighters to get the all the rope out of the rig, plus the spider straps and backboard.

"We don't know where the equipment is stored," whine the volunteers.

He commands them to find it anyway, and fast.

At the edge of the precipice, he directs the volunteers to tie a Bowline knot on the end of the backboard.

"What is a Bowline knot?"

McKane, now the teacher, shows them. One end of the rope is tied to a tree and the other to the backboard. Quickly now, the backboard is lowered to the injured parties below. Kaitlin is most injured. Alex and McKane wrap her in a blanket and carefully secure her to the backboard. Together, they hoist her to the ambulance above.

The paramedics arrive just as Kaitlin is placed in the ambulance. Then the sheriff arrives.

"You should all be dead," he says matter-of-factly.

* * *

As the ambulance is about to begin the transport, McKane stands at the open back door and watches Kaitlin and Alex. He sees Alex's impassioned look towards Kaitlin on the stretcher. He notices her ardent and tearful gaze back to Alex. Their reciprocal messages—wordless but undeniable.

McKane slowly but firmly shuts the ambulance doors, realizing that their life together is just beginning while a precious part of his own life is ending.

* * *

He catches a ride to the hospital to have the lacerations on his face, head and arms treated. The shoulder and back injuries can be treated later when he returns to the west side. He and his friends cheated death.

He should be happy, but he is feeling sad and confused. He wants to cry. He won't cry, certain of a deeper descent into the swirling abyss.

He calls his parents from the hospital and asks for a ride home. He explains the accident. He and Happy are okay, but her car is destroyed.

They both need to work on Monday.

On the return trip, there is silence. Silence, and a growing tension between them.

2

A New Direction

His job is firefighting. He has been at it for over two years if you count his training at the Academy.

You go on eight to fifteen calls a day. And each of those calls is the worst day of someone's life. It could be a drowning, heart attack, motor vehicle accident, electrocution, amputation, degloving incident or a fire. He learns how to emotionally detach just to get through the day.

McKane tries to shake off the accident. It's time to return to normal life. That thinking works up until the moment the flashbacks start.

Images in his mind haunt him. The flashbacks worsen whenever he sees an emergency response vehicle with activated lights and sirens. He talks to friends about how nervous he feels in the backseat of a vehicle. His friends are kind and sympathetic, but they can't know how terrible the accident was.

The flashbacks make it hard for him to work at the fire station. He begins to pass off his regularly scheduled hours to other firefighters who want the time. He mentions to the battalion chief that the accident had been intense. He's not doing well. The battalion chief listens to McKane, as he recounts the flashbacks and nightmares.

One dream involves taking an emergency call, arriving on the scene of a horrific accident, and then seeing himself dead in a ditch. He tells the chief he relives the accident whenever the station bell goes off. He fears that he'll make a mistake that could hurt someone. Safety is the heartbeat of the job—he can't afford to compromise anyone's safety.

The battalion chief is understanding and gives him one month off. The chief encourages him to get better.

"You are a good firefighter. We don't want to lose you."

One month later, McKane returns all his gear to the station.

* * *

McKane needs to replace lost income, so he turns his attention to building up the business he started a few years back, a pole-vaulting club called *ProVault NW*. The club occupies part of a large gymnastics' facility near the Boeing plant in Everett. He owns all the poles, pit, standards, runway, small tools, computers, and video equipment. There is a big whiteboard for teaching sessions, leaving messages, or doodling.

The facility is as good as or better than any like it in the country.

He teaches the pole vault with his own unique methods. His approach is novel and 180-degrees out of sync with traditional coaching. McKane invents teaching aids and builds novel training equipment to train budding athletes. Athletes from Texas to British Columbia fly in to ProVault to learn the skills he mastered while in high school.

As a high school freshman McKane's coaches first recognized his vaulting talent. They projected he would be state champion. To them he looked like a smaller version of Sergei Bubka, in the air but not on the runway. He glided, not jetted down the runway.

McKane set a state record of fifteen feet nine inches when only a sophomore. The next year he jumped sixteen feet eight and a half inches, easily winning the state championship. It helped that McKane had a background in gymnastics. Before pole vaulting, he had twice been all-around state gymnastics champion.

In McKane's vision, pole vaulting was just gymnastics on the end of a stick.

The clients at ProVault include athletes from age 12 to 82. They soon grow into a family of dedicated athletes, making the training sessions a blast. Sometimes a pizza party breaks out. In the summer, the group spends entire days pole vaulting in the sand along the Puget Sound. Crowds of spectators gather around loudly cheering successful jumps. Some of ProVault's clients enter national championships and win. Others earn college scholarships.

To McKane's satisfaction, his athletes begin to look up to him. They like him not just because he is an excellent teacher, but because he seeks to be every trainee's friend.

Some of the athletes develop closer ties to McKane. Greg, a high school senior with aptitude in many sports, is one such athlete. A football and soccer star, success on the field comes easy to Greg. His high school coach predicts he will be drafted into the NFL as a field goal specialist or maybe a fast linebacker. At the beginning of his senior year, Greg expands his athletic interests to include the pole vault. Within eight months of training sessions at ProVault, Greg is vaulting at big bars that should win him the title at the state meet.

Three weeks before the championship meet, however, Greg incurs injuries in a serious accident. A vehicle crashes into his car from behind. Greg's legs and back are aching, but he drags himself to vault practice. He finds he can't get back up on the big poles. Nevertheless, Greg competes at the state meet, playing through lingering injuries. He places a disappointing fourth. Not only is Greg's brief vaulting career over, but his entire athletic future disappears.

Greg enrolls in a university in California without the aid of an athletic scholarship, but he keeps in touch with McKane. Greg is hoping his vaulting coach can still serve as his life coach—and lifelong friend.

* * *

Eighty-two-year-old Butler is a "butler" in the true sense of the word, doing everything he can to serve everybody around him. Butler owns a company that builds parts for Boeing and for military aircraft. At age 80, he tries to retire but his wife won't let him. At age 82, he is still working.

McKane predicts he will probably die with his boots on.

Butler develops an exceptionally strong bond to McKane. They first meet in the summer at an "All-Comers" track meet in Edmonds. They share a birthday, December 4th.

Butler and McKane during a lunch break at the machine shop

Curiously, it is also the birthday of Sergei Bubka, the world's most famous pole vaulter—the athlete who broke the men's world record 35 times. Both McKane and Butler had learned to vault in their own backyards as children. Butler vaulted in high school and then later in the military. McKane vaulted in high school and then as a scholarship athlete at the University of Washington. Beyond a common birthday, they share a lifelong passion for the sport.

As their friendship grows, Butler discovers that McKane lost his four grandparents early in life. It is a loss McKane feels mostly around holidays. Butler being "butler" decides to appoint himself as McKane's honorary grandfather.

Thereafter, they make sure to meet or talk by phone every day.

It becomes habit for McKane to drop off lunch to Butler at the company shop, partly out of concern that Butler isn't eating enough. These shared lunches give the two of them structured time to talk about everything. Butler learns of McKane's secret dream to build a water-powered car. Butler relates to McKane's passion for invention, having developed hydrogen and oxygen splitting devices himself. In fact, he has been at it for a lifetime.

Having become acquainted with McKane's aspirations as an inventor, Butler allows McKane direct access to his plant and equipment. There McKane continues his work, taking advantage of the material inventory in Butler's plant. He learns to use the sophisticated machines and even assists the crew at the shop running the lathe.

Involving McKane in the production process is Butler's way of encouraging his surrogate grandson to follow his dreams.

* * *

At officially sanctioned track meets, McKane interacts with other vault coaches employed at high schools across the state, but never feels welcome or part of the coterie of school coaches.

He's a private coach—a businessman.

At the stadiums, the other coaches try to keep him off the field so he can't interact with the athletes. They make threats to call the police and have him arrested for trespassing on public property.

Some of the more aggressive coaches openly insult him in front of the athletes. One coach calls McKane a "piss ant" as he extends his hand in a mock friendly gesture at the state meet. McKane winces but keeps his head high in front of his athletes.

He's resolved not to surrender to the united opposition against him. He remains a constant presence at the track meets—and his athletes love him all the more for it.

* * *

Butler's favorite aphorism is "In the pole vault, an athlete can look so good or so bad and the difference is so little." Small differences in technique make all the difference. McKane uses the big white board at his facility to teach the science and the precision of the pole vault.

McKane's background in art helps him configure precise diagrams and illustrations. As a younger man, he had earned a master's degree in 3D Digital Art from the University of British Columbia, Simon Fraser University, Emily Carr University of Art and Design, and the British Columbia Institute of Technology. He also taught the Unity 3D Game Engine and Digital Art in Europe and Washington state for four years. For this reason, he feels comfortable and confident in front of a classroom, even without formal training in education.

Most days, the ProVault white board is covered with intriguing sketches with arrows, vector lines, symbols, and numbers. McKane comes to realize that even as a vault coach, he thinks like a mathematician—and an artist.

Sometimes after practice, McKane doodles at the white board, drawing illustrations of a theoretical hydrogen injector for his old BMW. His idea is to find a way to gain free energy with a new water splitting technique. McKane devises a simple, but novel electronic system to activate the release of hydrogen and draws it up on the white board.

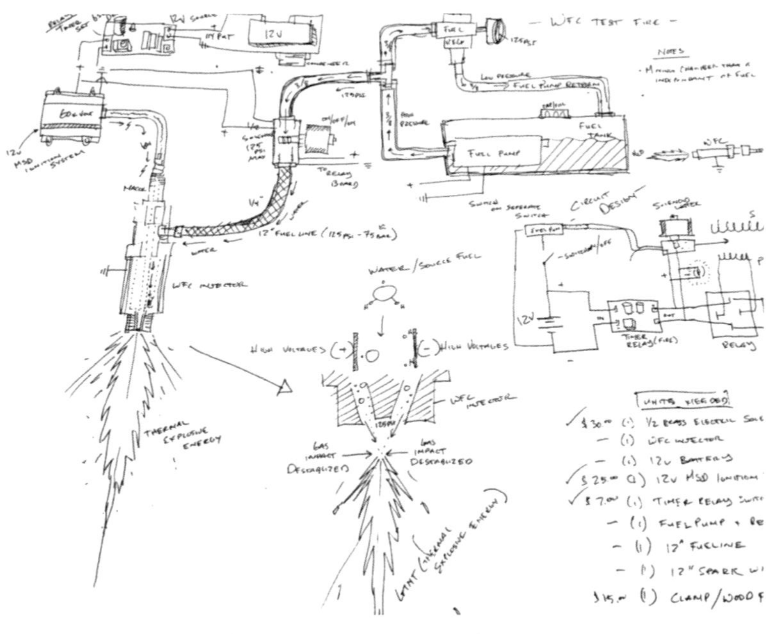

McKane’s drawing of the hydrogen injector system (2018)

On a few occasions, he leaves the after-hour sketches of his inventions. At the sight of them the athletes are fascinated and barrage McKane with questions about his theories. For no charge, he offers them a layman’s view of science, invention, spacecraft, clean energy, and fusion. McKane notices that the younger the athlete the more he or she intuitively understands the science he presents.

Nevertheless, some of the students' parents think he is a little crazy.

* * *

Life at ProVault comes to delight and comfort McKane. In his facility a group of wildly diverse athletes evolve into a loving, supportive community of friends. McKane embraces his organic role as their leader and enjoys his athletes' devotion and respect.

Nevertheless, in a close-knit circle of friends, it is hard to keep a secret.

Everyone at ProVault knows about McKane's accident—and its steep cost for him. Knowing his concern and dedication to each of them, his athletes are sympathetic and understanding in return.

Some even offer him hugs when he appears down.

After the accident athletes comfort McKane

3

A Necessary Detour

Battling a bout of depression for the first time in his life is difficult. It comes with a puzzling sadness that permeates everything. Losing himself in dysphoria feels like losing a portion of his soul, only worse.

McKane realizes he needs help to fully recover in the wake of his accident, and to reclaim what he has lost emotionally. His father, a clinical psychologist, refers McKane to a neuropsychologist famed for working with patients suffering with Post Traumatic Stress Disorder.

Dr. Burkhart's office is in Edmonds, a sprawling waterfront community with timber town roots. A Washington state ferry terminal is the city's beating heart. It is a gateway to Whidbey Island and the Olympic Mountains to the west, and the San Juan Islands and British Columbia to the north. The waterfront is lined with cafés, coffee shops, art galleries, a museum, a public marina, and a city park.

Edmonds is a beautiful city. The only problem is that the drive into Edmonds can sometimes be frustrating. Ferry rush hours feature congestion that rival I-5 traffic jams, long backups, and impatient commuters.

The first visit to Dr. Burkhart's office catches McKane off guard. He doesn't see the doctor as expected. Instead, the doctor's trusted assistant, Lee Farley, M.A., meets him there. Ms. Farley explains that McKane must undergo some psychological tests before seeing Dr. Burkhart.

McKane wonders, *what is a psychological test? Will I pass?* He concludes, *this is going to be a very serious, professional process.*

Ms. Farley asks him why he wants to see the doctor in the first place. In response McKane simply relates his memories of a car accident.

"I am a backseat passenger in a SUV. We are descending a steep mountain road. There is a snowstorm and a cliff . . . suddenly, we are flying. The car lands and begins tumbling. The car's roof is shrinking with each somersault. I push against the roof. It keeps buckling. Windows are crunching. My friends are screaming . . . I know we are going to die. I count the rollovers until the number twelve. I think, *this can't be happening. It must be a bad dream.* We survive. Now I'm here."

He mentions to her the flashbacks—how he's losing interest in golf and pole vaulting. How he feels nervous in the back seat of a car. How he's lost a little weight and his appetite.

How he's getting a divorce.

How he feels like a total failure.

McKane's had recurring nightmares about a ghost—a ghost who reminds him that he almost died without accomplishing his special purpose.

What that purpose is, the ghost doesn't tell him.

* * *

The first test is written. McKane reads the instructions. The page looks like a checklist. You just check a box if a listed problem is true for you. Sitting in the small interview room alone, he almost involuntarily checks the boxes "Depressed" and "Anxious."

He thinks, *what the hell am I doing with depression and anxiety*?

In the next test, McKane moves around puzzle pieces in a game called "Tower of Hanoi." Ms. Farley explains the task is to reorganize eight sizes of wooden disks, stacked largest to smallest in pyramid fashion, from a single peg on one side of the puzzle to a single peg on the opposite side of the puzzle. There is a central peg the player can use to accomplish the task as needed. Ms. Farley makes clear there are only two rules in the game. One is that only one disk can be moved at a time. The second rule is that a larger disk can never be placed on top of a smaller disk.

What Ms. Farley does not explain is that it takes 255 moves to correctly solve the puzzle.

McKane gets off to a good start. He easily moves five disks from one side of the puzzle to the other by shuttling the disks on and off the central peg. It's a good strategy. He might solve the puzzle in record time. Suddenly, he is stuck. He can't make another move without breaking Ms. Farley's rules.

His breathing tightens.

He senses the walls of the small exam room are shrinking. Thoughts embedded in panic accelerate. He feels frozen in space and time, trapped inside the puzzle. He is replicating his emotions in the collapsing Subaru.

Ms. Farley says he should stop for the day.

* * *

The first meeting with Dr. Burkhart does not go as expected. Dr. Burkhart presents as a normal guy, low-key and laid-back. He is anything but stuffy and over-intellectual, and his informal style sets

McKane at ease. The two of them are simply talking about the accident when the patient feels his defenses dropping.

Dr. Burkhart engages in talk therapy first. McKane talks. The doctor listens. Sometimes it is like a casual conversation with a close friend over morning coffee. Other times the dialogue turns more ominous—especially when Dr. Burkhart poses a piercing question.

McKane looks forward to the sessions. He feels better after talking about his work and things at home. He notices how Dr. Burkhart keeps circling conversations back to the accident itself. "What do you remember about the accident?" "What were your feelings right after?"

McKane feels blocked. He can describe the mechanics of the rollovers but not his feelings immediately after the incident. Dr. Burkhart gently but firmly persists. Calmly he repeats the questions, hoping to clinically evoke insight surrounding a latent memory. Instead, the doctor's questions only produce in McKane a rising anxiety.

"I must have been too busy with the rescue to notice," is McKane's half-hearted excuse for his lost emotional memory.

The doctor is not convinced. He senses his patient is still avoiding something—

Something important.

Something deeper.

Dr. Burkhart tries an experiment. McKane is sitting in a chair in front of a row of red lights. The lights are pulsing left to right. He stares at the lights without moving his head. When the lights stop moving, Dr. Burkhart directs him to close his eyes and remember the incident. The process repeats, each time with faster moving lights. On the fifth trial, Dr. Burkhart asks him to close his eyes and see himself at the crash site.

"What are you seeing?"

"I am at the edge of the mountain looking down as the car is flipping over and over. There is a ghost next to me watching. The ghost sees through the trees and through the snow. The ghost is in the trees and in the snow, watching. The ghost turns to whisper in my ear.

It's not your fault.

As these words exit his lips, McKane starts to weep. He doesn't know if the message is for him or Happy. It is probably for both.

"You are certainly taking to this well," Dr. Burkhart gently says.

* * *

McKane continues in therapy for more than a year. Sessions are every Thursday afternoon. Sometimes he forgets that he has severe PTSD. Dr. Burkhart doesn't remind him.

McKane comes to understand that he's not undergoing classical psychoanalysis. Dr. Burkhart's employs a unique approach to disentangling depressive patients from past emotional fixations. It's a positive process, gently opening the patient's insight to resolve hidden, intrapsychic conflicts.

McKane comes to accept that his past cannot be changed. And with time, he finds himself feeling more upbeat.

* * *

With one milestone achieved, Dr. Burkhart turns therapy in the direction of McKane's future. Like mending a fractured but invaluable work of art, the doctor understands the importance of post-traumatic growth as an essential factor in stabilizing recovery and finding one's "second life." In his view, successful psychotherapy propels the patient into a bright future built on the ruins of a broken history.

Dr. Burkhart encourages McKane to talk about his hobbies and what it is he would love to be doing for work five years from now.

"What is this thing you are spending all your time on at home?"

McKane explains that he is attempting to develop a water-powered car. Though progress is slow, McKane's eventual success is not out of reach. His principal challenge to success is a lack of time. He wishes his work on the car were his job instead of a weekend hobby. During the conversation, McKane confesses a desperate, if not irrational, urge to return to his hometown to finish his hydrogen-fueled car.

Having said it out loud, McKane is suddenly struck by an absolute conviction of what he must do next.

4

Into a Secret Forest

Six years in a garage.

The sentence is self-imposed. McKane chose to enter a virtual prison of four high walls and empty space, leaving everything else outside: wife, family, friends, vocation, and all avocations.

McKane is convinced his plan merits the weight of the sacrifice attached to his work. He's no longer a hobbyist. His aim is serious and significant—to develop a synthetic mechanism to trigger a water-to-energy conversion system. Based on an underpinning design he has been grinding out, his objective seems completely possible.

At least theoretically.

If he succeeds to plan, such a device would generate an earthquake in the world of energy production. It would be revolutionary. A technology like this could provide a blueprint to build practical, hydrogen-oxygen propelled machines of all shapes and sizes. A working model built from McKane's design could be tailored to fit any commercial or residential application.

And that's not all. A greater urgency compels McKane toward his goal. He's read and re-read the constant warnings from the scientific community about the catastrophes of a warming planet. Nevertheless, the vacuum of consensus on a plan from academia, political leaders, and economic minds to rescue the environment frustrates him. Urgency and anger fuel his single-minded pursuit of innovation.

McKane knows the climb ahead is steep. Given his limited credentials in science or technology, he has no business tackling the problem of climate change. In the back of his head, a voice on loop sneers at him.

You can do nothing about it, McKane. You are presumptuous to even think so. You are not a scientist nor even a credentialed educator.

McKane tries not to run the numbers in his busy mind, to calculate the odds against him.

Maybe the PTSD has made him crazy.

Some of his family and friends think so.

* * *

His dad and mom are firmly supportive. But that's not a mystery to McKane. They are tree-huggers from way back, now living on a small tree farm next to an Indian reservation.

McKane's parents grow six species of redwoods on their land, including the deciduous redwoods that once covered the entirety of North America a million years ago. They even raise Red Dawn redwoods, once thought to be extinct. The only evidence of the tree's existence could be found in the fossil record until 1943, when botanists visiting a remote region in China unexpectedly found a valley of the huge trees. This landmark discovery led to a resurgence of the species in Europe, the United States, and his parent's property.

McKane's parents understand the power of scientific discovery, and the stakes of their son's work. They offer to stake him financially, even as McKane throws his own savings at the project.

* * *

McKane stows his equipment and carries out his work in a large garage space on loan from a kind widow. The makeshift laboratory sits on a solitary plot somewhere in a forest north of Seattle. Though he can't afford a long-term lease of such a large workshop, McKane does chores around the woman's house and acreage to make up the difference.

But rent money doesn't interest her anyway. After all, she's retired with a good pension from Alaska Airlines and owns two other garages. The shop she loans to McKane was her husband's special retreat, a massive 'mancave,' with workspace and storage for his prized RV and custom-built fishing boat. According to his surviving wife, the man often tinkered there with lights flooding late into the night. He had lived a satisfying and long life full of adventure.

In the widow's tales of her husband's life, McKane senses an echo of his own ambitious spirit.

Just after high school, the man had learned to fly bush planes across the Alaskan landscape. Decades later he took a job as a pilot for Alaska Airlines out of Seattle. There he met his wife, a chief stewardess for the airline. Years into his retirement, the man held onto his flying credentials. Though he never lost the hunger to fly, his legs eventually lost the strength to carry him up and down the airplane ramps.

After retirement, he often went fishing alone in the ocean. Ultimately, he died from health complications after a boating accident. The widow's husband had fallen into the frigid waters near Sitka and was unable to extract himself for three hours.

Though she rarely enters the garage now serving as McKane's workshop, the old woman finds comfort seeing the lights in the big shop as McKane works late into the night. It reminds her of better, more loving times. With no immediate family around her, she welcomes him into her world.

* * *

McKane's lab fills up one side of the garage. The garage offers him more than enough room for his old BMW, a computer, various small tools, and machines. He places a cache of old science books at the end of the workbench. *Light and Matter, Catastrophe Theory, Radiotron, Modern Dictionary of Electronics, Physics of the Impossible, and McGraw-Hill Concise Encyclopedia of Science and Technology* are his go-to references guiding his journey. Like the alchemists of old, McKane somehow feels there are magical secrets buried in scientific antiquity. He would not have discovered his passion to be an inventor without these old books.

It is never easy finding the very old books. Many are lost. Others are forgotten after rejection by the scientific edifice of the day. But McKane's library is built on a stroke of luck.

Rena, one of ProVault's Masters National Champions, calls him one night while he is studying in the lab. Her father, a theoretical physicist, had just died. Like McKane, Rena's father had a life-long passion for science—specifically, an interest the study of gravity. Rena's father believed very few people understood the mystery and power of gravity. And he had left behind an expansive library: a treasure trove of old books, technical notes, and manuscripts.

Rena offers the resources to McKane, who thanks her profusely.

Day and night McKane works alone in his self-styled monastery. The days turn into weeks, months, and years. He comes to believe that without a fixed daily schedule he will relapse into PTSD.

His father, the psychologist, gently echoes this concern.

* * *

He takes only coffee in the morning and then a light snack or sandwich in the afternoon. Dinner is taken out or a quick meal at his parent's house. At night, he reads. Anything having to do with hydrogen density and gravity. He devours audio books about quantum field theory and physics. He finds old, expired patents that inspire him.

McKane is ready to take the next step. The problem is, McKane lacks the academic qualifications to complete proper scientific research.

* * *

McKane has no formal training in the hard sciences. He took some biology, chemistry, and physics classes in high school. But those were more introductory surveys rather than rigorous science. And his grades were average. In art classes where he was allowed the freedom to use his imagination to create something new McKane thrived. Most of all, he excelled his peers in drafting class where the focus was converting mental impressions into precise forms, diagrams, and illustrations.

At the University of Washington McKane had elected art as a major. This path allowed him the flexibility in his class schedule to be in the track pavilion twice a day. He embraced the grueling rhythm of weight training sessions in the morning and vaulting practice in the afternoon. Every Division One scholarship athlete understands the delicate balance of managing a full-time "job" on the field while at the same time dealing with all the academic requirements.

And if your GPA drops, you get fired.

McKane enrolled in all the core art classes. He found himself gravitating to figure drawing and print making, earning a reputation as an excellent sketch artist. The drawings consist mostly of fictional creatures, battles with futuristic weapons and armors, and imaginary power reactors for spacecraft.

Given the flexibility in his academic course load for elective classes, McKane picked science courses rather than the lightweight classes others took to bump their GPAs. Physical science intrigued him more than anything else. The trouble was formal departmental approval was required to enroll in upper division science courses. This became a constant hurdle for the young art student.

McKane signed up for a 200-level Mechanical Engineering 3D Design class and passed with flying colors. He discovered the existence of another 3D Design class—a 400-level course for Mechanical Engineer majors only. He petitioned the department for a waiver to get a seat in the class, based on his interest in computer 3D coupled and his prior success in the 200-level class. The course focused on CATIA—a digital content system frequently used by professional engineers. CATIA was the industry standard, used at Boeing and throughout the aerospace field.

At first, McKane adopted the role of observer. Mainly, he listened to the lectures and avoided chatter with his classmates, having little in common with the engineering majors. For their part, they politely tolerated him.

But he overheard the sarcastic comments about the "art major" in the room. An academic paranoia soon set in. In the back of his mind, he remembered hearing from a wise man, *only the paranoid survives*.

To his engineering classmates' surprise, McKane made the Dean's list that semester.

* * *

McKane moved on to courses in physics and astronomy, where he earned solid *A*'s. On one occasion, the *Physics of Light* professor invited him to her office. For a while, they conversed in turn about the Hubble telescope, new-found galaxies, meteors, meteoroids, and vehicles needed for future space exploration. Before he left her office, the professor said that McKane inspired her.

5

Discovery

He approaches lab work as a painter to a canvas—with an open mind patiently awaiting the illumination of creative energy. It is his biggest asset. Invention, after all, is a product of the imagination first.

McKane doesn't feel gifted. Rather, at the bottom of his psyche, he feels himself to be a unique species of human: half artist – half inventor, half man, and half ghost. Since the accident, he senses having one foot in three-dimensional reality and the other across a forbidden, invisible boundary in space and time.

He is not sure on which side of the boundary he belongs, or even which he prefers.

His soul is bifurcated in a way he can't quite explain.

* * *

The day begins inauspiciously. He is in the lab working on the same old problem: how to efficiently split water into its constituent parts. His prime project for a thousand days. He has no special plan for the day ahead. There are no insights flooding in to guide his action. He understands having no destination at all is the perfect state of mind for work like this.

He is experimenting with a new configuration of his electronic board and its interface, crafting what he thinks of as a “resonant cavity.” But he is frustrated. His novel electronic scheme isn’t working as planned.

McKane turns up the voltage. The amps spike up and the voltage drops. Nothing happens in the water. He can find resonant frequency with the coils and the water, but no hydrogen is produced. He tries one hundred different configurations.

Nothing works.

He pushes harder, his efforts lasting the day. Finally, he steps away from his equipment, totally exhausted.

McKane sits down and rests his head in his hands, granting himself brief reprieve from desperate impulses. Unwanted thoughts begin to intrude, preoccupying his mind and reshaping his mood. Soon, a stark realization hits him.

This is it: the end of the journey. The insights were wrong. The project has failed. It was a terrible idea. Years lost forever. There is nothing ahead.

In McKane’s psyche, already fragile thoughts and feelings enter freefall.

He tries to put his mind back in a neutral gear, focusing on controlled breathing to detach from the swirling vortex of desperate thought. He waits, watching for the moment to pass. It doesn’t. Instead, a no-warning flashback catapults him into a dormant memory.

It is his last panic attack years ago in the psychologist’s office—the time when he couldn’t move the wooden doughnut onto the right peg—trapped within the boundaries of an impossible puzzle. Now, he finds himself tangled in another enigmatic maze, with far more ominous consequences.

McKane is crying.

PTSD is in control, propelling his thoughts, reminding him *everything is pointless.*

He wonders if he should just kill himself and be done with it.

Suddenly, another vision comes to him, more powerful than the last.

Through his tears, McKane sees the ghost from the mountain. The ghost is in the lab, watching him. Once, the wise apparition stood with him at the edge of an alpine precipice. Now it has returned to be at his side on the edge a deeper abyss.

The ghost sees the machine. The ghost is in the machine. McKane hears a faint whisper.

Just do it like this.

McKane understands the ghost's command, but he cannot obey. It makes no sense. No electrical engineer would configure electronics the way he suddenly sees it in his head.

It could never work. He comes close to ignoring the ethereal admonition and giving up.

* * *

The inventor stands and wipes the remaining tear from his cheek. Disentangled from doubt, he moves in blind obedience to the soldering station to begin reconfiguring his electrical board exactly as directed.

McKane connects the new setup to the resonant cavity and energizes the system.

The device begins to pulsate with energy. The voltage is skyrocketing even though the amperage is extremely low and stable. Something strange is happening. Masses of excited hydrogen and oxygen bubbles are powerfully surging through the system.

Astonished, he wonders, *where is all this energy coming from? I've never seen this before. No one has ever seen this before.*

His eyes pass back and forth over the machine as revelation forms in his brain.

I can synthetically produce off-the-chart levels of combustible, pristine hydrogen, and oxygen gas at only 1.9 watts.

He lingers in the lab, staring at the unbelievable energy readings coming from the miniature machine. Moments later, he is at the white board doing the math. It works out. The electrical input is small—only 1.9 watts—while the energy output is powering two 5-watt light bulbs.

The weight of the discovery begins to dawn on his tired mind. The very existence on the benchtop invention changes everything.

He cannot begin to fathom the implications of the discovery.

* * *

McKane feels as though he just witnessed a birth. He wants to celebrate the event, to share the cheery moment surrounded by loved ones and colleagues.

But there is no one with him in the lab and hasn't been for five years.

A sharp pang of loneliness travels through McKane's body like a muscle spasm. It awakens a memory of another time when he should have been happy for cheating death on the mountain, but instead he was struck down by depression.

He moves to shut off the machine, then the lights. He closes and locks the door to the lab. Under the dark sky McKane traces his steps down a worn forest path, the quiet corridor to his tiny living quarters, where no one is waiting for him. With a sudden headache seemingly out of nowhere, he drops heavily onto his bed and falls asleep as if nothing happened.

* * *

McKane wakes early the next morning. He lingers for a moment on the edge of the bed and tries to recall every detail of the day before. A question keeps revolving in his mind: *where is all this energy coming from*? He agonizes over the steps necessary to answer the question. An enigmatic puzzle in science may have been solved.

He wonders if he can replicate the proof.

McKane returns to the lab even before morning coffee He opens the door and steps through slowly—deliberately—like a man entering a religious cathedral. The soaring ceiling still consists of sheet rock and white paint, not large, imposing stone with ornate borders of imported tiling. Yet, it feels like he is entering the Catholic church where his sister, Dejah, attends services. He reverently approaches the newborn object on the workbench as if it were a historic relic of great significance.

He gathers up the forensic evidence still inside the machine, at first hesitating to touch the object for fear of disturbing the exact wiring configuration that produced the magical release of excited oxygen and hydrogen gas.

Would it be a sacrilege, he thinks, *if the lost Benben stone from the top of the pyramid was discovered, and I dared to touch it*?

The water is still in the cylindrical resonant cavity at center of the device. He calls it the *well*, and later he will call it something else more ominous. He sees the residual water in the well has turned a yellowish color. Before the test, it was clear well water. As if handling a priceless ancient sculpture, he carefully extracts most of the treated liquid with a pipette vial bottle and tightly secures the cap.

But now McKane notices hundreds of black particulates inhabiting the remaining liquid at the bottom of the well. He wonders, *what are these nuggets and how did they get there*? He will need to capture the remaining water and the black particles together. McKane grabs a second pipette vile and carefully evacuates the resonant chamber, planning to segregate the sediment with a soft microfilter and then set it aside to dry before further analysis. But first McKane grabs a neodymium magnet to see if the rock-like material reacts to the presence of a magnetic field. He slides the magnet around the outside of the glass vile.

The black particles start to move and then to swirl in the watery matrix, precisely following the movement of the magnet.

Then another impossible truth dawns on him.

The invention can synthetically produce magnetic rock from pure drinking water at only 1.9 watts and at room temperature.

He formulates a plan to get third party verification of the mysterious discovery.

6

Following Quantum Breadcrumbs

McKane walks through the front door of Edge Analytical, a highly accredited environmental testing laboratory in Washington.

He has no appointment.

Politely, McKane tells the receptionist he wants to see the manager. She talks to someone on the conference line passing the message, then turns back to him. As they wait, McKane makes conversation with the young lady, who tells him, “It is nice to meet such a positive person. Most the people that come in here are kind of grumpy. They just want to get their stuff and get out.”

A tall gentleman in a lab jacket enters the reception area, greeting McKane with a big smile. His name is Bryce. He wants to know how he can help.

“I want to have a water specimen tested. I have it here in this glass vial.”

Bryce takes the vial of treated water from McKane and holds it up to the ceiling light. He says, “Oh my god, that’s a lot of iron in there!” Bryce asks where it came from.

"I made it. I have a proprietary process. It is like electrolysis only 180 degrees different."

McKane realizes that what he just said will not make sense to the Bryce. Or anyone else, for that matter.

Bryce continues, "Are you putting anything into the water before energizing the system?"

"Nope. It starts with either well water or seawater. That's it."

Interested, Bryce pledges to obtain a full list of elements from McKane's sample, including the heavy particles from the golden water.

* * *

One week later, Bryce calls to report that the data values on the sample are in. McKane asks for the details.

Bryce explains that the energy process increased the iron content in the well water by 3,405 per cent when compared to the untreated water sample. He points out that this change is far more than expected from standard methods of electrolysis, even though stainless-steel electrodes are composed mostly of iron. Even more surprising, McKane's experiment only ran a few hours, not for days or weeks as often happens in other electrolysis' methods. Most startling to Bryce is that the experimental process did not require emersion of the electrodes into a dielectric spiked with KOH—potassium hydroxide—which facilitates the ionic exchange in the water. McKane's untreated source looks like unaltered local well water or pristine seawater without any expensive chemical boosters. Bryce points out that nitrogen and silica typically decrease in standard electrolysis, but these elements increased substantially in McKane's experiment.

Bryce concludes, "Whatever is going on inside your device, it's a special rearrangement process."

* * *

Next, McKane establishes email contact with Werner Kaminski, who leads up X-Ray Crystallography at the University of Washington. They agree to meet in person at his office on campus to discuss the black, magnetic sediment left in the resonant chamber after the breakthrough discovery.

It is a day in late August when McKane arrives on the UW campus. He's 15 minutes early to his appointment with the professor. Early arrivals are McKane's habit since training in the Fire Academy. "If you are on-time, you are 15-minutes late," was the Firehouse rule.

Old habits die hard for the former firefighter.

McKane enters the Chemistry building to find the hallways empty. The fall quarter hasn't begun. For a moment, he imagines these halls heavily populated with chemistry majors engaged in muffled conversations about experimental processes like his. Today the halls are quiet. Each step on the tile floor echoes a hundred feet.

McKane's wearing dress shoes to match his dark suit and tie. He's not quite sure why he dressed up. As an artist turned inventor, McKane is new at this.

He guesses the correct path to Dr. Kaminski's office, as there are no directional signs. At the end of the main hall, he takes a right turn, then a left. McKane discovers a number on a blond wooden door and figures he's reached the end of his search.

The door is locked. No one answers his knock. He wonders if this is the right door or even the right building for that matter. McKane waits in solitude as quiet minutes slowly pass. It seems an age until his attention is captured by the sound of a door opening nearby and then loudly slamming shut.

A middle-aged gentleman wearing casual clothes is cautiously approaching. His gaze is fixed upon McKane as if he were an intruder.

"Hello there. Are you from the government?"

McKane is confused by the salutation. *Is the professor serious? Or is it simply a joke to keep the situation light?*

With a smile, McKane glances down at his formal attire and suddenly understands the professor's first impression. He assures the professor he is not a federal emissary.

Dr. Kaminski unlocks the door to his office. With his hand still on the doorknob, he says, "You must be McKane. Come on into my office. Would you like a cup of coffee?"

The professor offers McKane a chair. The two are now facing each other in a cramped office filled with papers, articles, beakers, textbooks, and a computer, in no particular order. As the coffee is brewing, McKane notices the walls are covered with academic awards, commemorations of scientific accomplishments and memorabilia from former students.

Dr. Kaminski stands to open a window. It is going to be a warmer day than expected in Seattle. He turns back and asks, "Why are you wearing a black suit and tie? Are you sure you are not from the government?"

McKane realizes he missed the punchline of the doctor's earlier playful tease. He laughs. "No, I am not a secret agent. I used to be a student here. I graduated awhile back with an Art degree. Sometimes, I just like to put my best foot forward—that's all."

Something about this man suggests to McKane that the professor can be trusted, that the two of them may even one day be friends.

Dr. Kaminski pours a hot cup of coffee for him and asks, "Now, how can I help you today?"

McKane explains that he is experimenting with a new type of electrolysis. As the professor patiently listens, McKane walks him through his process. He's dissociating water molecules into their hydrogen and oxygen parts without using chemicals or any chemical equipment. His dielectric can be well water, sea water or even ambient air. McKane describes his use of low levels of electrical input, not even enough to turn on a lightbulb. In this experimental process, he couples a centrally positioned water chamber, or "well," to a novel electronic board. The result is water molecules splitting apart profusely and at room temperature. McKane shares his hunch that the energy produced by this process could easily be harvested if he had the right equipment. Finally, he mentions the curious black sediment that is magnetic, and suggests that the sediment might be magnetite or even a synthetic meteoroid. He hopes Dr. Kaminsky can help identify the newly formed rocks.

"Did you bring the samples with you today?"

"Yes, of course." Slowly and deliberately, McKane passes Dr. Kaminski the specimen across the table, as if he were handling a small, exotic creature still sleeping in its glass cage. Dr. Kaminski examines the subject. "Wow, this looks crystalline."

Black Magnetic Rocks – Magnetite (Fe3O4)

McKane produces a neodymium magnet from his black computer bag and hands it to the professor. “Look, if you take the magnet to the glass bottle, the rocks will follow it.”

To the professor’s surprise, the rocks obediently follow the small magnet, racing along the inside edge of the bottle to keep up with the movement of his hand.

In the space of a few moments, Dr. Kaminski’s mood shifts from playful to serious. He stands up and directs McKane to follow him. The professor opens the door and motions with his head toward the hallway. Leaving their full coffee mugs behind, they start down the still vacant hallway.

Dr. Kaminski enters a large room inhabited by electronic devises of all shapes and sizes. It appears to be the Chemistry Department’s main laboratory. The former art-student-turned-inventor follows, his imagination spinning.

What I couldn’t do with all these magnificent toys!

The professor is at the lab table preparing the mysterious specimens for analysis. He first examines the sample with a light microscope.

“Look at these shapes - they are already crystalline!”

As he continues, McKane notices the excitement and urgency mingled in his voice.

“They have sharp edges. This is very cool.”

Falling silent, Dr. Kaminski examines the object further. “It appears we are looking at a black body that deflects light, not absorbs it.”

Next, Dr. Kaminski places the tiny rock on a sticky thimble head and into a mounting tray. He carefully loads it for X-Ray Crystallography. He explains what should be seen on the screen and why.

"The monochromatic x-ray beam hits the specimen, promoting and producing a diffraction pattern of rings. These ring formations tell us what the specimen is."

McKane feels himself nearly trembling with anticipation. It is the feeling of a parent watching an ultrasound screen about to reveal the identity of a child. The inventor does not move.

Unconscious of it, McKane holds his breath as he watches the machine's screen.

As the ring formations begin to appear, something suddenly goes wrong. The image on the screen isn't forming properly.

Dr. Kaminski stops the test.

"Sorry. I'll have to tinker a bit more to get an accurate reading for you. It seems the machine is having difficulty reading your specimen."

The professor pleads for a continuance of a week or more to get some more answers. McKane grants him the time. *Good things come to those who wait,* he thinks.

* * *

A week goes by. It seems longer. Dr. Kaminsky finally contacts him, sending an email message with an attached image. McKane clicks on the attachment first. The screen opens to reveal a beautiful set of rings, organized, and striated much like rings of a tree, or even the rings of Saturn. He reads Dr. Kaminski's message.

> *It's fascinating, but I do not completely understand what the ring formations mean. This is a very complicated specimen, perhaps the first of its kind. It would take me months to determine what this is. You may want to check with the Molecular Analysis Facility (MAF) to get a better understanding. There is a nice lady named Samantha Young*

there that does XRD (X-Ray Powder Diffraction). She can help you identify the sediments.

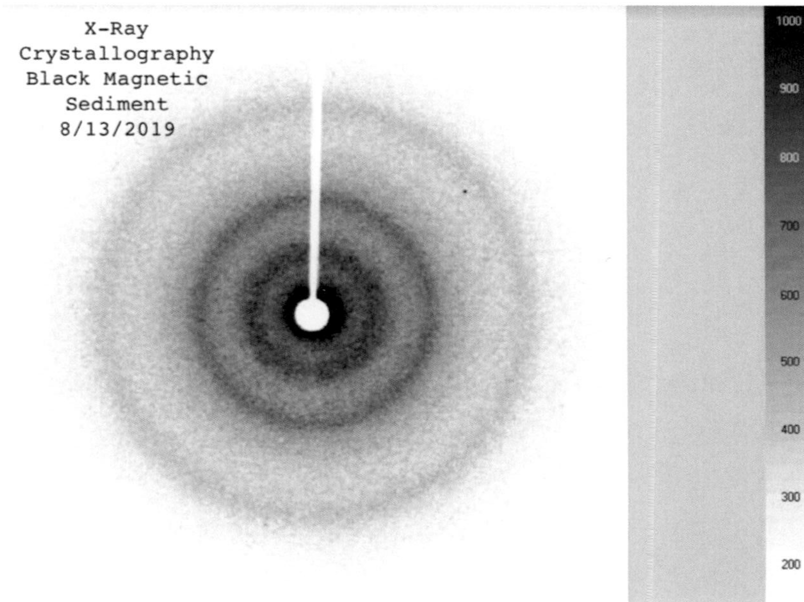

Kaminsky's X-Ray Crystallography Photo

A few days later, McKane drops off a sealed pouch of Kauai Coffee at the blond wooden door. It is, as McKane discovered in their brief conversation, Dr. Kaminski's favorite.

McKane feels like he is entering a fold of great scientific colleagues with each step forward in his journey.

7

Into Quantum Space

He has been in the lobby now for ten minutes awaiting entrance to the Molecular Analysis Facility.

The MAF, as it is known, occupies the ground floor of the Molecular Engineering and Science Institute building at the center of the University of Washington campus. It has a reputation as the largest vibration-free lab space on the West coast. He thinks the building should also receive recognition as a majestic work of modern art—a beautiful working space enveloped in engineered sheets of glass. Inside, it is big, open, and organic.

Entrance to the MAF is by a secret passage. You must be accompanied by an internal UW official wearing a security badge. The badge activates the elevator to the MAF, housed one story down. Students and visitors can take the same elevator to higher floors but not to the lower level without observing established security protocols. He realizes the MAF is not on the ground but *in* the ground.

He hasn't forgotten Dr. Kaminski's directive to see Samantha Young after the X-Ray Crystallography failed to reveal the truth about the miniature black stones. McKane is carrying a black briefcase in his hand. Inside are specimens that may be of great consequence. He must find out.

McKane is dressed more appropriately for the occasion today, less formal, so as not to repeat the initial misunderstanding with Dr. Kaminski. While waiting, McKane notices a young lady enter the open lobby. She walks straight toward him and smiles as if they were longtime friends. In reality, they have never met in person. But McKane had sent the woman an introductory email in the second week of September.

> *Greetings Samantha - my name is McKane Lee. I'm an alumnus from the University of Washington, way back in 2007. Yikes, I know, weird that after all these years; I'm back on campus searching for answers to life... Lol. I've been trying to get some samples analyzed here at the UW. It has proven quite difficult based on the samples I have synthesized. Martin sent me to Werner, and then, Werner sent me to Tatyana. And now, after meeting Tatyana today, she is directing me to you. It's wonderful, I get to meet everyone! And I must say, everyone has been wonderful! I'd love to meet up and discuss the process and see if these four samples can be properly analyzed with the Bruker D-8 (tight focus range) in your department. I postulate that the samples will be small pits of black/brown Fe3O4 and Werner found another white substrate (not visible with the naked eye) but was unable to make out the X-Ray Diffraction scattering pattern... Werner mentioned potentially calcite... There are just so many questions because the black flakes are also strongly attracted to a magnet. I know school is about to start and "things" are getting crazy busy on campus. When would be your first available time to meet? I'm free and I care very much about getting to the bottom of this puzzling question or at least get started... Hope this email finds you well. Thanks for your precious time. Best.*

Samantha Young, PhD, is a Materials Chemist and the Instrument Manager at the MAF. Understanding her background gives McKane additional confidence. She works with outside industrial R&D applications and alongside the top crust academic peers solving complex measurement problems. Most importantly, Dr. Young has access to the most sophisticated electronic instruments in the world. The MAF facility is state-of-the art.

She wouldn't waste her time with McKane. There are thousands of chemists around the country she could consult with, if needed. But she believes there is something to McKane's scientific experimentation. He feels reassured that Dr. Young will work through any challenges posed by analysis of the remnant elements.

Dr. Young's office is also underground. The two of them walk together towards the elevator adjacent to the lobby. Dr. Young activates the elevator with her security badge, and they step in. Moments later the elevator opens to reveal MAF.

The entirety of it is large and appealing, modern in design, pristine and sanitized. McKane thinks briefly that this place rivals the stories told about Area 51. He shoos away the thought, but it's a challenge. The reality of this space is more impressive than the imagination.

Dr. Young's office is just beyond the elevator door exit, a glass enclosure surrounded by many other glass enclosures. She leads him past her office to a larger room, a laboratory. Along the walls are electronic machines arranged in meticulous order. Desks, tables, chairs, and shelves have been arranged strategically so as not to interfere with the room's openness.

Dr. Young greets a colleague with a badge that reads *Scott Braswell.* She introduces Scott to McKane, mentioning that McKane is doing research on water and energy. McKane makes a quick mental note that Mr. Braswell works with electron microscopes.

After the brief introduction, they cross through the lab until they are facing two large XRD machines side by side against the wall. She selects one of them, indicating this is the machine that will analyze the specimens. It works, according to Dr. Young, by scanning back and forth over the top of the samples. Significant crystalline peaks reveal information about the composition of the subject material. She assures McKane that this machine should reveal the identity and composition of the materials.

Her confidence gives him hope.

* * *

A week goes by before McKane receives an email message from Dr. Young. Soon after, he receives another—followed by yet another. She reports having trouble understanding the samples, especially the distilled water paired with copper electrodes. According to Dr. Young's review of the machine results, the water has iron in it when it shouldn't. And the particles are magnetic when they should be inert. Though she could make educated guesses, Dr. Young is unable to describe the identity of the black particles without more information about its crystalline structures.

She has questions for him about the origins of the material.

To McKane her words reveal a sense of bewilderment, almost frustration. And her next words seem to close the door on further inquiry. "I will return your samples to you", she says. "Anytime next week if you would like them."

McKane is crestfallen. After visits to three world class laboratories, the scientists he encounters are genuinely fascinated, but frustrated and devoid of answers. Their instruments cannot establish the identity of the sediment that appeared in a small machine sitting on a workbench in his makeshift garage-turned-laboratory in the woods.

McKane tries to comfort himself with the ancient aphorism of Confucius.

It doesn't matter how long your journey takes as long as you don't stop.

I can still do this. I must solve this, he thinks, nursing the same feeling he felt after "failing" a puzzle in a psychologist's office years ago.

Perhaps the truth of it lies farther below the surface.

* * *

McKane opens his computer, gets on-line, and goes to the MAF home page.

After rummaging through the index of names and associated specialties, he stumbles across a name he remembers: "Scott Braswell." He had almost forgotten Dr. Young's introduction of Scott on the day of McKane's first MAF visit. According to his professional profile, Scott Braswell works with the most advanced version of the Scanning Electron Microscope.

Just what McKane needs to get answers. He quickly composes an email.

> *Hello Scott Braswell. My name is McKane Lee and I'm an alumnus of the University of Washington way back in 2007. I've recently been conducting some XRD sampling data with Samantha Young. She mentioned your name and I'd like to inquire about potentially snapping a few pictures with your electron microscope. I've got two electrodes both measuring the same dimensions. The dimensions are 4" long and 0.50" diameter. They have developed a strange coating on them. Samantha and I have trying to determine*

the atomic structure. Samantha is writing up a report at this moment on her findings with XRD.

I'd like to get high resolution visible documentation on the material on the electrodes. Would you be able to assist with this? I'm not interested in being trained on the equipment. I'm in the middle of research and development and don't have time to allocate to the machines. Nor would I want to accidentally screw something up on such an expensive apparatus.

Please, let me know when you would like to touch base and meet up. I know school is about to start soon. I've got to visit the department regardless to pick up the source sample from Samantha. So, I'd like to hit two birds with one stone.

Would you be available tomorrow? Hope you are well. And I am very much looking forward to chatting.

Best,

McKane B. Lee

P.S. I've attached two pictures of the samples.

Having introduced his purpose, McKane waits for a response. Several days pass without a reply, and McKane wonders if there will even be a response. *Should he resend the message or make a phone call?* He senses he is on an urgent quest, but time is short, and it's difficult to be patient.

McKane braces for another disappointment.

A message from Scott Braswell finally appears in his email inbox, featuring a personal invite back to MAF. Scott reminds McKane to bring the samples, so he can get started right away.

McKane's mood soars. This time, he hopes to find a consequential truth.

Days later, McKane enters the multiplex through double glass doors engraved with the UW / MAF emblem. He turns right into the imposing lobby, where Scott is waiting to greet him and to lead the way underground. McKane is relieved to be back at MAF.

In truth, he has nowhere else to go now.

***

They walk together into a glass-encased room with tables, chairs, lab equipment, manuals, and the Scanning Electron Microscope (SEM). The two of them sit down side-by-side at a table facing the SEM. McKane hands Scott the dried samples of what appears to be tiny, copper-colored nuggets that he had harvested during a recent test run with well water and copper electrodes at 1.9 watts for four hours.

Scott prepares a sample and slides it into the SEM to perform the microscopy. He scans the object to select a target of interest. After locating one in the image, he increases the magnification.

“Look at the crystals! They are all facing the same way!” he gasps. “I can get even closer. This is something very special, amazing! I have never seen anything like this before!”

Turning to McKane, he asks, “What are you doing to get this?”

Copper-Oxide Quasi-Crystal Formations

"I am here for you to tell me."

After a moment of silence between them, McKane stands up, walks to a whiteboard, and begins drawing a figure. He creates a graph with a x-y axis. He inserts numbers on the graph and symbols on the board. The rudimentary drawing signifies magnitudes of mass, hydrogen densities and molecular binding energies. He stops and lays down the marking pen.

"There."

"Oh my god, you are doing transmutation." Scott says.

"Your words, not mine."

Scott continues, "This is so cool. If you have any other samples, bring them to me at once!"

McKane thinks, *Scott is sure taking to this well.*

On the long I-5 commute home, McKane notices the traffic ahead feels lighter than usual.

8

Partners Into the Portal

Scott Braswell and McKane partner for the next two years. McKane runs the tests in his lab, then brings the results to Scott's lab. The process never changes.

Scott uses the SEM to analyze the sediment under super high magnification and then reports back. The photos are out-of-this-world. Images never seen before. As though under superstition, the two of them never openly discuss the implications of being on the edge of a significant discovery.

They don't want to spoil the magic of discovery by saying it out loud.

* * *

Throughout their extended partnership as researchers, Scott has never seen McKane's actual tabletop device—even a photo of it. He is only interested in analyzing the molecular products of the process. Though unspoken, the two of them share an understanding that they will limit talk about the underpinning technology. Whatever it is, it already exists, sitting on a garage workbench in a forest. Their focus now is compiling and unifying a body of observations of pre-test and post-test water samples derived from the Quantum Kinetic Well ("QKW™").

McKane anguishes over proving the science of it, even though the findings are reassuring, consistent and repeatable. In every experimental trial, the observation is that atomic elements are transmuting into an array of heavier elements. The originally sourced particles in the water seem to fall apart, or are pulled apart, while migrating to the bottom of the QKW™. At touch down, the elements have new atomic identities. Magnetite, silver, aluminum, yttrium iron, niobium, zirconium, iron, and other novel elements consistently appear as if from no discernable source.

Pen-rose octagon quasi-crystal tiling

McKane repeats the pre-test and post-test procedures time and time again. And Scott confirms the results.

Scott is staggered by another observation arising from his research. While the QKW™ effect produces many new elements, it also mitigates concentrations of certain noxious elements. The post-test samples consistently reveal markedly lower levels of deuterium, tritium, strontium, and salt. Even the electrodes are changing, sometimes growing in weight—not eroding away as in traditional Faraday electrolysis.

More breathtaking still, these transformations begin at only 1.9 watts of power. There are no artificial additions of thermal power or heavily spiked chemicals to escalate the process. Only natural well water, ocean water or distilled water are used. And the device operates perfectly at room temperature.

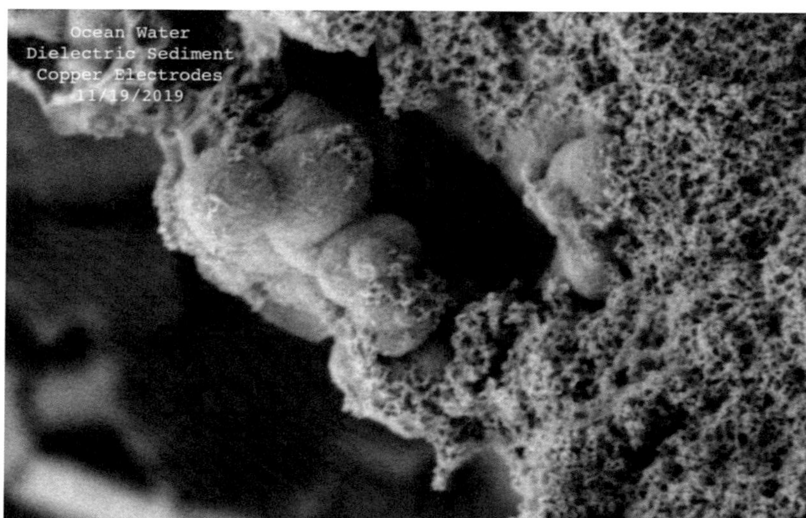

Paramagnetic nano-sphere(s) sediment

Scott can't get his well-trained scientific mind around the emerging realization that anyone who can turn on a light switch can now turn on a transmutation machine. The only difference is that it takes less energy to activate the QKW™ than to power a tiny Christmas tree light bulb.

Scott and McKane avoid another subject. They're actively testing a theory currently held in disfavor by the scientific community.

Once considered the holy grail of scientific inquiry for more than 40 years, the theory often described by scientists as "cold fusion"—that electrolysis conducted at room temperature might yield a significant energy source—was abandoned forever after decades of disappointed attempts in the laboratory. The official death certificate of the concept was issued in an editorial by *Nature* magazine a few years earlier.

Once holy ground in scientific fields thereafter became forbidden territory to any scientist wishing to preserve his standing among peers or academia.

But McKane will not accept conventional thinking on this subject. Nor can he disregard what he and Scott have now looked at under the most powerful microscope in the world—the molecular structures of previously unknown objects. Seeing new elements emerge in a quantum of space and time is nearly stupefying—a world of synthetic space rock and organic first forms. The images are both shocking and beautiful, like something lifted from a horror movie.

Or the first chapter of Genesis.

Hollow Carbon Nano-Sphere

* * *

Every few weeks, McKane had taken the journey to the MAF to deliver new samples to Scott. The drive took 55 minutes in good traffic, but up to two hours at rush hour, because of Boeing traffic out of Everett early in the afternoon.

McKane would like to meet regularly with Scott, but Covid-19 restrictions now bar his access to MAF. So, McKane begins leaving the samples for Scott in a locked box inside the lobby.

Nevertheless, the inventor and partner scientist continue communication by text or email and seldom talk on the phone. Though now confined to separate labs, they nevertheless share a sense of purpose.

Alone in his lab again, McKane feels himself to be a de facto secret agent, working on a secret mission. The problem is that no government agency, or even academia, acknowledges him or his work. Disavowal comes in many forms, from explicit rejection to the cumulative burden of economic deprivation. He hasn't had a paycheck for more than five years.

But it helps to work with a well-known and skilled scientist at the MAF, one who understands the gravity of the work they're doing.

On April 21 of 2021, Scott reassures McKane in a facetious note: "I am always available to look at more synthetic space rock from some alien tech out of the seas of Atlantis for you. I can reserve some microscope time on Monday if you like."

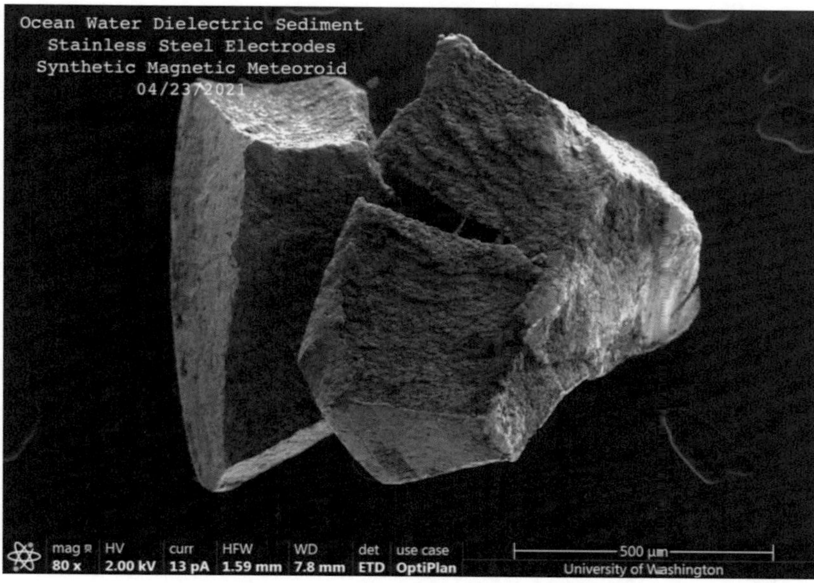

Hyper magnetic synthetic black rocks (CFe3O4)

9

Unexpected Vistas

Contaminated water.

McKane wants to introduce the world to a future without it.

By additional research McKane comes to suspect that his invention opens the doorway to low-energy commercial water purification plants hundreds of times more powerful than existing water treatment systems. He envisions clean, cost-effective methods to remove large quantities of radioactive isotopes from spent nuclear wastewater. If McKane is right, small modular versions of the QKW™ could be installed in undeveloped geographical areas to ensure pure drinking water. Desalinization systems could even be simpler and far more proficient, using his technology. Clean energy sources would now rely on nature's own free-energy conversions systems.

And based on the strange, almost unreal findings in his lab, he believes all these breakthroughs are possible in the time span of a single lifetime.

Too much evidence is the right amount, he reasons. This is never truer than when standing in the courtroom of public opinion—and when the jury pool will consist of academicians and scientists from across the entire planet. McKane realizes that the honesty and justice of his proof will soon stand trial in the court of public scientific opinion. And he cannot shake his worry that a rigid scientific community will reach an unfavorable verdict, despite the facts on his side.

With concern about a skeptical reception of his work growing in his consciousness, he thinks of another of his father's humorous quips: *When in doubt, overreact!* So, McKane embraces his father's quip as advice. He reaches out to a national water expert before a worldwide trial.

* * *

Gerald Pollack is a larger-than-life Biochemist at the University of Washington. McKane has been impressed with Dr. Pollack's theories on the science of water ever since he read his books *The Fourth Phase of Water* and *Germs and Gels and the Engines of Life*. He is aware that Dr. Pollack was just voted as one of the top one hundred "Most Inspirational Persons in the World" by the OOOM 100. Along with most of the scientific world, he has viewed Dr. Pollack's educational videos on YouTube and TED Talks.

McKane sends an introductory email to Dr. Pollack.

> *Greetings Jerry,*
>
> *My name is McKane Lee, an alumnus from the University of Washington (2007). I'd like to meet with you sometime next week if possible.*
>
> *I've been engaged with the Molecular Analytical Facility (Nano-Technology Department) using the electron microscope and have found some amazing discoveries I'd like to share with you. I strongly believe it is right up your alley in terms of water. To be more specific, the "4th Phase of Water." Or is that the Quantum Phase of Water? I'm sure we are talking about the same thing... right?*

In the long run, if you are interested, I'd like your support of a patent pending device. It might be worth your time. Want to meet and learn more?

Best,

P.S. Sneak peek: picture attached.

McKane B. Lee

To his surprise, McKane's inquiry prompts a quick response from Dr. Pollack. They set a day to meet at *Zoka*, a popular coffee shop in the University District near Laurelhurst, at 10:30 a.m.

McKane arrives too early. On purpose. He wants a little time to rehearse his presentation of key SEM images of the hybrid structures. McKane scouts the venue and selects a table at the back of the café, one perfect for discussing topics of great importance. Then, he orders coffee and a blueberry muffin, which sits untouched on his plate.

There's too much excitement for him to eat anything.

Several minutes past their agreed meet time of 10:30, McKane scans the room again, looking for Dr. Pollack. He worries that his morning will end in disappointment.

At 10:50 a.m., McKane stands and makes his way to the front of the café, a weathered computer case dangling from one hand and the uneaten blueberry muffin in the other. He takes one last look around the room. His eyes suddenly latch on the famous scientist sitting alone at a table near the front door. Dr. Pollack is pensively finishing a muffin. His coffee cup is empty.

"Hello. You must be Dr. Pollack," McKane says, stepping up to the professor's table. "I recognize you from your web page at Pollack Laboratories."

"Lucky you came by. I was just about to leave."

“So sorry, doctor,” McKane says as he takes a seat. He offers his untouched blueberry muffin to the professor, as if to compensate him for the loss of his valuable time.

Dr. Pollack graciously accepts. It is his favorite.

As rehearsed, McKane opens his computer to begin his 50-slide Power Point presentation. The introductory slide is a cross-section of planet earth, including a red hot, molten core.

Dr. Pollack stops him before he can advance to the next slide.

“What makes you think the core of the earth is hot? Has anyone been there? How do we know it is not ice cold? The first rule of science is to ‘be careful about what you assume is generally accepted knowledge.’”

Chastened, McKane replies, “That makes sense. Magnets become demagnetized when super-heated. If the core of the earth was molten hot, then the center of the earth would be demagnetized. It is more likely the earth’s core is a gigantic chunk of magnetized space rock, magnetite. You are right, we won’t know until we get there.”

The professor simply smiles.

McKane gets his presentation back on track, advancing to slides with selected images from the SEM analysis. He shows the professor photos of the Quantum Well, the resonant cavity where all the magic happens.

Dr. Pollack stops him again.

“Wait a minute. I know what you are doing. You are doing low voltage transmutation, aren’t you?”

McKane makes no unnecessary admissions. Instead, he says, “Funny you say that. Scott Braswell over at the MAF said the same thing.”

“I wish I knew how to do that. I have been trying to do that for a long time.”

McKane humbly accepts the professor’s comment as a great compliment.

Their conversation turns to recent UW research projects, to their unique family dynamics, and to personal catastrophes they have both experienced.

Dr. Pollack describes how the pharmaceutical industry has been opposing his research on dynamic water technologies. McKane, echoing the sentiment, explains how a group of engineers have repeatedly scoffed at his research on water and energy.

After two hours, Dr. Pollack announces he must get to another appointment. "I wish we could talk longer. It's been great. We are now officially friends."

McKane has no words, only a sense of gratitude and hope.

The professor continues, "You know, they say I am the 89th most inspirational person in the world. Today, you have inspired me."

10

On the Edge of a Quantum Matrix

Soon, Dr. Pollack sends McKane a message relevant to the future of the technology and how to bring it to the world.

"You need to contact my friend Matt. He is leading a team of top scientists working on fusion and quantum computers. Mention my name to him and you will get a response," Dr. Pollack promises, hoping to fashion a path forward for his new friend.

McKane comes to learn that Matt is the Program Manager for a large research project revisiting the theory of cold fusion, the idea that experimental, modular nuclear fusion machines can be established to imitate the power of the stars, except at room temperatures. The long-term strategy is to build such devices if the technology can be established scientifically.

So far, no one has.

McKane realizes that someone like Matt might just be the type of visionary needed to propel the novel transmutation technology forward. He composes another message of introduction.

Meeting to Discuss: "A New Quantum Technology".

Greetings Matt. My name is McKane Lee. I'm an inventor and patent writer for Quantum Kinetics Corp. Our research team has been utilizing the University of Washington for testing, sampling, and evaluations of a "New Quantum Device". The reason for my letter: I recently met with Gerald Pollack at the UW over coffee last week. I showed him the device and the application(s) of the device, and he immediately recommended I speak with you ASAP. Would you be able to meet up sometime soon for a presentation of our patent pending quantum device? Please, let me know. I've also attached a little Christmas "Image" for you in this message. Enjoy! I sincerely hope you had a wonderful Christmas and have a Happy New Year! This is going to be a "Historic" year for technological discovery for humanity! Talk soon.

Best

McKane B. Lee

P.S. "Thoroughly conscious ignorance is the prelude to every real advance in science."

James Clerk Maxwell

Matt and McKane first meet online. It is in the first days of the worldwide pandemic, and the research team executives are now working from home, cautious about Covid-19.

After breaking the ice with a comment about how they might be wearing the same blue shirt, Matt gets right to the point.

"What is it you are doing?"

McKane describes his decade-long interest in water and energy. It began with designing and building a water-splitting hydrogen booster to improve his car's gas mileage. To achieve this technology, he was able to dissociate water molecules into hydrogen and oxygen using generally accepted chemical methods. But those methods seemed to be inefficient and costly. Then one day, he discovered a novel approach completely out of phase with traditional Faraday electrolysis. It was a physical approach to electrolysis based on voltage, not amperage, to dissociate the dielectric medium. In his lab, this novel process produced unreal amounts of energy at only 1.9 watts. In addition, it transmuted molecular structures into new elements. The machine has a small nuclear signature, possibly low energy, super-soft luminous x-rays.

"I call the machine a Quantum Kinetic Well," McKane finishes.

"It has everything we look for," Matt says, matter-of-factly. "The challenge is how to establish your commercial propositions."

McKane asks what he means.

Matt describes the process. As he explains, an inventor like McKane must first establish intellectual property rights. Once he has a patent utility, then he can talk about his work openly and begin to license it to industry. Different industries will use it in different ways. Early applications of the tech will be crude. The founder of the technology will need to spend time working with industry to develop uses and applications. Don't get involved with those who wish to suppress the science. Open Pandora's box. Put the science on the table. Let the marketplace evolve around the new technology.

Matt advises McKane to be sure he has others around him committed to the process—the more, the better, so it can't be shut down.

Matt further explains his belief that humans have evolved based on proof of concepts that challenge people's sense of order. The stereotyped image in the public's eye is that "experts" should be the ones offering the proof and telling the story.

"What you are doing will challenge the stereotype."

Matt continues. "Invention is hard; making a product is harder. To start, you need to make a full narrative about your journey as an inventor. You need to convince people and get them excited about a new technology. There is a value in *writing it down*. Expect people to be incredulous and dismissive of you. You won't be the first."

McKane is taken off guard by Matt's frank vision about what is about to come in the next five years on planet earth. But he feels compelled to follow Matt further down the rabbit hole.

"What you need to know is that there is a big ocean swell about to break. There is new research. Everyone who has been briefed about it knows it is coming. It will explode like a supernova. You need to ride the lightning."

Spellbound, McKane crosses with Matt into an imaginative theoretical *wonderland*.

"We need to have 100 top minds to be working on the research together. You need to decide whether you want to be part of a collaborative network or work alone."

In addition, Matt gives a warning: "Despite your significant discovery, a 'go-it alone approach' will not work. Small inventors get stuck when they are under-capitalized."

The first hour of discussion passes quickly. Matt has an urgent matter to attend to. They agree to continue the conference in 45 minutes.

* * *

McKane finds himself catching his breath. This meeting was far more consequential than even he anticipated. As he waits for Matt to return to the computer screen, McKane decides he has time for a second cup of coffee.

His imagination has been stoked by Matt's bold prognosis of imminent breakthroughs in technology. McKane lets himself wonder, *What's next?*

Matt returns to the computer conference. Their meeting was originally scheduled to end after an hour. *As a Program Manager at such a prestigious, internationally known organization,* McKane thinks, *Matt's time must be extremely valuable. Matt must have something more on his mind.*

When their meeting reconvenes, McKane notices a different tone and tenor to their conversation. The teacher-student dyad is over—it has been for an hour. Matt is now talking to him as if they were already travelers on the same path. Matt describes his past experiences with research on cold fusion and the failures of the scientific community to prove the theory. He describes the current research on quantum computers, and technical challenges encountered. Matt wonders out loud if the quantum computer will be able to teach the world how to build cold fusion machines.

McKane offers him a riddle. "Have you ever thought that the 'cold fusion' device *is* the 'quantum computer' you are looking for?"

Matt hesitates a moment before saying, "I never thought of that."

"Remember, Matt, the shortest distance between two points, A and B, is not a straight line, it is a singularity."

***

After the meeting, McKane concludes Matt is a modern visionary, and as history proves, great prophets are often unappreciated in their own time.

McKane's machine can do many wonderful things. The inventor is convinced of that more than ever. And now he has a blueprint to bring his invention to the world. Matt has handed him the key, by way of powerful advice, to unlock many doors on the way forward.

As always, McKane walks through the next portal, not knowing if peril or promise lies beyond.

McKane wonders how it will feel to live in Wonderland.

11

Maxwell's Demon

McKane reaches out to a national laboratory in Colorado to ask for assistance in analyzing the energy capacities of his invention.

This time, in keeping with Matt's advice, he makes a full disclosure on the front end of his request. He's testing a new form of electrolysis, one that produces abundant amounts of hydrogen and oxygen fuel at only 1.9 watts. The process relies on voltage, not amperage, to dissociate the water molecules. After a month of promising discussions with scientists and administrators, who initially seem quite enthusiastic about the research, the stream of calls and emails from the national laboratory suddenly dries up, without explanation.

Undeterred, McKane reaches out to a large laboratory in Texas, making the same pitch. "We don't know who you are," the lab's representative says bluntly.

Thereafter, no one at the Texas laboratory will return his calls or emails.

* * *

Facing several slamming doors to his appeals for laboratory help, McKane begins discussions with a scientific team at a university in

British Columbia, the same university where he attended two years of graduate school. They express a measure of interest in the project. However, they express concern about not having the current "bandwidth" to undertake such a study. Plus, with Covid-19 restrictions still in place at the Canadian American border, it is not possible for him to establish a working relationship in Canada.

McKane makes a presentation to a large group of engineers in Seattle, setting forth his theories and diagrams of the quantum machine. He lays out his bold vision regarding future applications of a new technology.

Before he even finishes, the professionals are making sarcastic comments to the artist turned inventor. The gallery makes catcalls, and someone even shouts-out "bullshit."

But McKane knew this was coming. He had been duly warned, by both Matt and Dr. Pollack. Still, the defamatory comments ring hot in his ears.

McKane reaches out to a good friend and faithful ally, an immensely intelligent woman. He asks for her counsel at an inflexion point in his life, reminding her he values her opinion above all others.

"Should I keep going?" he asks.

When she responds, he can't believe her words.

"You should have never started this project. It is impossible. It's been too painful for you. You have given up everything. You should think about going back to school now and start over before it is too late."

Her vote of no confidence lands hard on his ego. Repeated disappointments drag on McKane's confidence.

He fears he's slipping back into a loop of obscurity and defeat.

* * *

McKane strikes up a dialogue via email with an astrophysicist at a prominent university, hoping to engage the esteemed scientist in a discussion about solar fusion, the formation of cold bodies in space, and the chemical composition of meteoroids. When McKane offers a cursory overview of his research and the mysterious discovery in his garage-turned-laboratory, the professor initially seems interested. But when McKane turns to discuss a potential new form of plasma energy, the physicist quickly dismisses the matter as "nonsense"—and McKane's work along with it.

McKane consults with another contact over lunch at *Anthony's* in Edmonds. The man works as a scientist at a nuclear facility in eastern Washington. McKane gives an abbreviated version of the speech he made to the group of Seattle engineers.

Inadvertently, he uses the word "Bremsstrahlung," a German word for electron braking radiation.

The nuclear engineer says, "What is "Bremsstrahlung?"

"Don't you know?"

The nuclear engineer gets out his phone and looks it up. "Oh, yes, now I remember reading about that in a textbook."

The engineer looks back at McKane.

"If what you are telling me is true, and you figured it out, then I am sitting across the table from the richest man in the entire world. What is your secret?"

McKane replies with a story.

"Do you remember the famous thought experiment by James Clerk Maxwell? It is a parable about an opening doorway between two compartments full of gas particles. There is a tiny, imaginary creature who sits at the door. Every time the creature sees a fast-moving particle approaching from the left-hand side, it opens the door and lets it into the right-hand compartment. And every time a slow-moving particle approaches from the right, the creature lets it into the left-hand compartment. Pretty soon, the left-hand compartment is full of slow, cold particles, and the right-hand compartment grows hot. So, what Maxwell did was create a puzzle within a paradox, a mechanism that defies one of the laws of the thermodynamics, entropy."

"Okay, what's the point?" asks the engineer.

"What if the imaginary creature controlling the portal is a basic electronic switch?"

"Bullshit."

McKane picks up the bill and pays the gratuity on his way out of the restaurant.

12

Overcoming Barriers

It takes him several days to comprehend what the incredulity of a seasoned nuclear scientist, an astrophysicist, the enclave of Seattle engineers, and two regional laboratories all add up to.

McKane is an unknown inventor and an uncredentialed scientist, and as such he will never hold their respect. McKane may have made a marvelous discovery, but his novel technology will never be accepted by the scientific establishment if it is *his* voice alone advancing a new form of energy.

* * *

McKane knows what it's like to be shunned by your superiors, even if you have remarkable skills of consequence.

He remembers his emergence onto the pole-vaulting scene when he was an unknown high school freshman competing at the District Championship Meet.

His competitors in the finals were juniors and senior students with established athletic backgrounds, school records, and names affiliated with years of feted accomplishments in track and field. All were competing for invitation to the State Championship scheduled in two weeks. Only the top three would qualify.

He recalls the glory of that day—and the pain. He remembers waking too early on that May morning, sparkling sunshine flooding his room from around the edges of an opaque window shade. That sun told him all he needed to know—the afternoon air would be warm in the Stanwood stadium. There would be little if any headwind from the southeast, and even if the wind picked up from the northwest, it would only propel him faster down the runway, leading to higher heights.

In his memory, McKane sits on the edge of his bed, unwilling to move further until visualizing once more his well-practiced technique on the runway and the strange gymnastic moves he will make in midair. He replays Sergei Bubka's powerful takeoff in his mind's eye.

Already, he is feeling slightly nauseous. He won't touch the gourmet breakfast kindly prepared by his father.

Before McKane leaves to catch the school bus at the end of the long country driveway, his father reminds him there is a chance that the UPS might deliver the new, stronger UCS Spirit pole today, sometime before the meet. McKane had spent all his savings to purchase the pole. His dad will bring it to the stadium if it arrives in time for McKane to use it.

At school, the freshman boy is distracted. He cannot concentrate in first period French class, despite Ms. Weingarten's daily effervescence. He only passively engages with his new project during Art class. He may be only fifteen years old, but he has other, more momentous things on his mind. Inexperienced and naïve, he dreams of local immortality.

All the Arlington track athletes with qualifying marks will be automatically excused from afternoon classes. The team bus is scheduled to leave campus at one o'clock. McKane elects the early lunch period in the school cafeteria. There is no time for socializing or happy chatter. He has little appetite, still consumed by nervous energy.

Before boarding the bus, he feels a strong compulsion to empty himself of the sparse remnants of lunch. He does not do so, the threat of public humiliation too great.

There are hundreds of people inhabiting the Stanwood stadium. It is filled to the rafters. Some are his friends, others family. Mostly, these onlookers are strangers who have never heard of the newcomer. His father is there, along with his older sister, Amorah—a track star in her own day at Arlington High School. McKane's mother is not present. She is working her regular schedule at the hospital in the L&D department, where she cannot be excused during a scheduled delivery.

The pole vault runway is positioned squarely in front of the stadium seating. Twelve vaulters have already been eliminated from the competition, missing at various heights on three consecutive attempts. The crowd is boisterous and sit on the edge of their seats, having already been thrilled at amazing vaults by veteran vaulters—and the one novice.

Along with McKane, only two other vaulters remain in the competition. One is his good buddy from Stanwood High School. The other is the event favorite, a muscular star quarterback on the football team and former two-time winner of the district vault competition.

McKane slips off his gray and blue warm-ups and stands at the end of the runway. He is falling behind in the competition, having more misses at lower heights than his two remaining rivals.

His name is announced over the stadium speakers, giving him just two minutes to complete his last attempt. The bar is set at what would be McKane's new personal record. In his hands is the new, stronger Spirit pole he has never used, the one dropped off by his father. The UPS driver had arrived with the package just in time before the meet.

McKane is taking a huge risk, not knowing if he can handle the longer pole without a single practice attempt.

His focus is the crossbar, but McKane is also keenly aware of his audience. He turns to face the crowd, waving his right arm over his head, pleading the crowd for more energy. He senses they are with him, invested in an underdog's last chance at victory.

He is an athlete but enjoys the theatre of it all.

His run is powerful and purposeful. He presses the pole above his shoulders and leans slightly forward into its strength, initiating a powerful pole plant. The precise and infinitesimally quick movements of his body in the air leave the crowd in a frenzy as he flies two feet over the crossbar.

Moments before clearing the bar...

His gutsiest move ever has paid off. Only his father and his co-competitor from the Stanwood squad understand what has just happened and what risks he took to triumph in this way.

The remaining vaulters fail at the height, elevating McKane to champion.

But that is not enough glory for the day. McKane decides to up the ante. He races over to the officials and asks them to set the bar even higher. Another foot higher.

The once clamorous stadium falls eerily quiet, as if they are witnessing something of historic significance. A young talent, previously unknown to them, is bursting onto an athletic field for the first time.

Once again, McKane plays to the crowd, clapping his hands above his head, inviting their combined energy to assist on his last attempt. The bar is set well beyond his prior record. When he clears the bar with ease in picture-perfect form, the crowd again erupts. He turns to acknowledge them with two raised arms, generously sharing his victory with everyone present.

Out of youthful enthusiasm or inexperience, he gives a final salute and then performs a celebratory back flip off the back of the vault pit, a type of gymnastics seldom seen in such an arena. The result is a crescendo of approval from the stands.

With congratulatory athletes surrounding him, lifting him high on their shoulders, the Arlington head coach and the team's senior captain push through the throng to confront McKane. Their unfriendly faces are matched with a hostile message. Openly, the coach and team captain confront him, saying that he may be an amazing athlete, but he has embarrassed the entire City of Arlington with his showmanship and arrogant stunt off the back of the mat.

Confused and embarrassed by the rebuke, McKane gathers his equipment and quietly leaves the stadium. He returns home in the family Suburban, instead of taking the team bus back to the campus.

* * *

A month later, in a crowded high school auditorium, the Arlington track team is introduced. They receive ecstatic praise from the entire student body. The team's many accomplishments and records are reported, including their impressive performances at the State Championship in Tacoma.

But no mention is made of the notorious young vaulter or of his legendary jumps in the Stanwood stadium.

13

Into the Nation's Laboratory

Pained by a sense of rejection from the mainstream scientific community, McKane determines that he must tune out all voices opposing the new technology. To move forward with any success, he must surround himself with open-minded, supportive people.

He chats with Scott Braswell at the MAF about his frustrations in getting third party verification of his machine. Scott suggests reaching out to Pacific Northwest National Laboratory ("PNNL"), a national scientific facility and subsidiary of the Department of Energy. Scott says the PNNL campus is close by in Sequim, only a ferry ride and modest drive away from Seattle. To Scott's knowledge, the University of Washington and PNNL have a long and established relationship. Many of the scientists at PNNL either attended the UW as students or worked at the university as professors.

Scott provides an unlisted PNNL phone number to call, promising he will vouch for him whenever needed.

The same day, McKane calls the number. No one answers. An anonymous voice invites him to leave a message.

Hello. My name is McKane Lee. Scott Braswell over at the UW gave me this number. I have been doing some research with electrolysis that is being studied at the MAF. I need to get some numbers and feedback about the process. I live close by. Please let me know.

Two weeks pass with no answer.

McKane is convinced there won't be one.

While brainstorming an alternate source of verification, however, he receives a call from a young lady with a European accent. She indicates that she is calling from the national laboratory. For two weeks, she has been trying to place his project with the most appropriate agency.

"You need to contact the Oceanography Department. They will get you going. They know you will be reaching out to them directly."

The woman shares the number for the scientist at the national oceanographic laboratory chosen to lead McKane's research project. The facility is located on the shores of the Strait of Juan de Fuca in the Pacific Northwest.

* * *

They first meet by video conference. The lead on the project appears to be standing on the deck of a fishing boat in the middle of the Pacific Ocean on a beautiful summer day. It is an inviting scene to McKane, who has fished for salmon off the Washington coast on his father's boat since he was a boy.

"So, I assume you like fishing?"

"Yeah, especially for Coho salmon out here in the Strait," says the scientist.

"Me too."

"So, what is your project?"

McKane explains that he is working with a new type of electrolysis. He's trying to nail down what is happening inside the device. He references the energy production of the technology, and the curious sediment the electrolysis leaves behind. McKane wants to know about changes in the isotopes of the water. He suspects the synthetic device can efficiently alter atomic isotopes with little energy input.

But he wants the details laid out in a scientific publication, if possible.

"Can you do that?"

"Oh yes, our lab can definitely do that, no problem!"

The next time they meet, McKane finds himself standing in the John Wayne marina, near the entrance to the research enclave. Because of Covid-19 restrictions, non-authorized persons are not permitted into the sprawling federal facilities. The facility is fringed with tall fences, barbed wire, and ominous warning signs forbidding entry.

The team leader and McKane work out an approach to the project, agreeing to work as strategic partners to compile and unify observations of pre-test and post-test water samples derived from McKane's Quantum Kinetic Well.

They set up a paradigm for a series of experiments. The Leader obtains water samples for each test, using two sources for the dielectric: well water from an outlying area and seawater from the Strait of Juan de Fuca, the purest in the world. McKane meets the project Leader up at the Marina parking lot every month for a year. The scientist delivers large containers of fresh and salt water in sanitized carboys along with pristine equipment to collect the treated samples.

McKane then runs the tests in his lab and returns the treated water samples.

At that point, lab personnel, all with advanced degrees in oceanography or engineering, use the most sophisticated instruments on planet earth to study both the treated and untreated samples. The Lead marshals the evidence and collates the data into summary form. As a true experimentalist, he simply describes the tests and reports the results, without speculating on the underlying chemistry, biology or electronics producing the results. This way the findings should be unassailable.

The body of observations compiled by the national laboratory prove to be exceptional.

On first impression, the data overwhelms McKane. Complex and highly extensive, the data will require more than a single journal article to relate.

It may require a lifetime of journal articles.

He decides to limit the first publications to the exotic findings most likely to redress the most urgent environmental challenges: clean energy, universal pure water, and carbon sequestration from the atmosphere.

14

Fusion Unto the World

Inspired by the first round of scientific tests, tests which confirm the invention's capabilities to manipulate atomic isotopes at a low energy, McKane takes his invention to the United States Patent and Trademark Office ("USPTO") and Canadian Intellectual Property Office ("CIPO") at the same time.

He hopes to put a timestamp on the science of his machine in North America as a starting point. After writing and filing a lengthy "White Paper" describing the concepts implicit in the new technology, McKane submits official applications for utility patents. In the patents, McKane discloses the invention's design and function, even providing diagrams on how to build the machine. He lists thirteen formal claims about its operational capacities and forecasts the far-reaching applications of the device.

McKane sits and waits. Time seems an enemy, interminably postponing fulfillment of a dream. He realizes it may take two to three years to complete the patent review process, or to even get a visit from a patent officer investigating the invention. Many patent applications are denied. Other inventors go to Federal Claims Court to plead the validity of their inventions. Some utility devices are inexplicably sequestered, like Hollywood's version of the "Ark of the Covenant" in *Raiders of the Lost Ark*, buried forever in some secret vault for national security reasons.

Five months go by. It becomes difficult for McKane to keep hope alive, given the odds against him.

While on a brief visit to Kauai, he receives an unexpected phone call from Ohio. It is his patent lawyer, leaving a voicemail.

> *This is Roger Emerson. I'm calling at 1:30 on Friday the 24th, that's 1:30 Eastern Time. I just want to tell you just got word about five minutes ago from the U.S Patent and Trademark office that they are approving the patent application we filed on McKane's Quantum Kinetic Well. So, that is exceptionally good news and normally we would have another six-or-nine-month fight on this, but they approved it right away, which I think is a testimony to how different McKane's ideas are than what's out there. So, I want you to know that.*
>
> *I know you recently sent me an email asking the status on things. Before I could even check on the status, this came in. So, your timing was great. We just got word a half hour ago from the USPTO. I just got the word only five minutes ago. So, congratulations. We are sending out our normal reporting letter along with the form letter that gives you all the details. But I just wanted to let you know right away.*

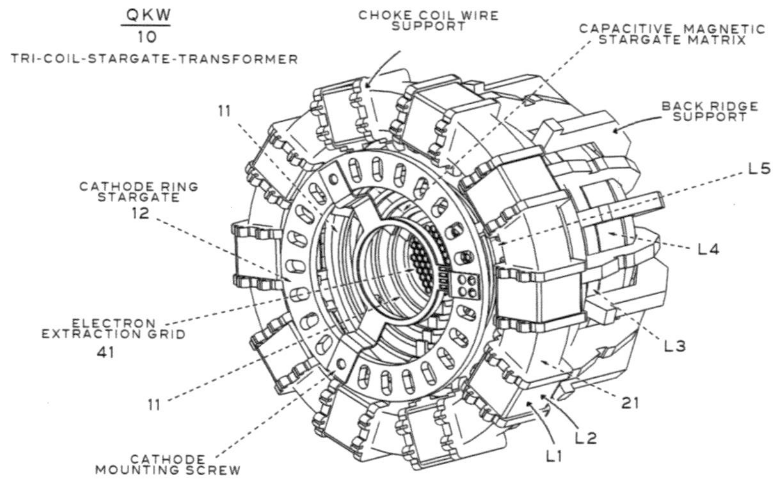

The patented Quantum Kinetic Well (Arc Reactor)

McKane cuts his vacation short and sets up a phone call with his lawyer to garner more details.

As his lawyer relates, the patent officer issued the patent quickly because it was clear McKane's device "worked" and that there existed "no other devices" the same-as, or substantially similar-to, the subject invention. The patent officer expressed personal consternation that scientists had not yet discovered this technology.

McKane receives the official paperwork approving his patent. It is overwhelming. He asks his patent lawyer if there is any further verbal communication with the patent officer. There is. It comes in the form of a personal request, or perhaps even a prayer.

Go teach this to the world.

He contemplates the significance of the unofficial message. He reads the official patent award letter over and over, absorbing its significance. *Could it be that a synthetic marvel that opens a domain of boundless clean energy has appeared just in time to save humanity's home planet?* The magnitude of the commission, and its urgency, suddenly strikes home with McKane.

There is no time, or need, to wait for decades of further scientific research and analysis. The great world powers are facing off against each other to see who will capture the last drop of ancient oil. Emergency use authorization is required to stop the madness.

He reasons it can happen with a new energy source at hand.

McKane calls Roger Emerson, his lawyer, again. He wants to file for additional patents in Great Britain, France, Australia, New Zealand, Canada, and the United States, as a start.

McKane catches Roger up on a recent success in his own lab with another invention that has been in the works for a year. The patent narrative for this device is already drafted and ready to go. He explains that the new device can be coupled to the QKW™ to greatly increase its power. As an injector system, it can be applied as a retrofit technology for use in internal combustion engines, commercial airline jets, heating furnaces and rocket propulsion thrusters. The injector system utilizes hydrogen, the perfect renewable fuel for all industries because there is a natural liquid storehouse of hydrogen–water.

McKane points out that the combined thermal explosive force of hydrogen and oxygen is astronomical. That is why most existing rocket thrusters rely on cryogenic oxygen (LOX) and cryogenic hydrogen (LH2) (solid fuel) as the fuel source within a combustion chamber. But McKane's injector fuel system is sourced differently—it's not reliant on cryogenic oxygen and cryogenic hydrogen. Instead, the device harnesses the ferocious atomic-power yield from room temperature water (water-mist as fuel), vectored for rockets thrusters or any internal combustion engine on demand.

Roger agrees to start work on the injector patent. As he does, the Super Lawyer receives an important message. It is from the Canadian Intellectual Property Office. McKane's application for patent of the Quantum Kinetic Fusor, the fusion device itself, is coming through. Approval of the invention comes with a familiar message from the examiner: *The invention works. There is no competing technology. All claims are allowed.*

Roger updates McKane on the good news. The inventor, and his lawyer, are again startled by the swiftness of the award. Precedent in patent approval suggests it could have taken years to gain approval, if ever. McKane's expectation of court battles and governmental sequestrations dissipate into thin air.

Examining the patent paperwork, the inventor notes a striking similarity between the Canadian patent officer's message on the QK Fuser to the United States patent officer's message on the QK Well.

Different countries . . . same message.

With a flush of humility, McKane considers the prodigious implications of this second patent. *It is now possible to couple the patented QK Fusor with the patented QK Well, together fortifying the gateway to nature's abundant energy domain.*

On the heels of the CIPO revelation, McKane receives another game-changing message from his colleague at the national oceanographic laboratory.

Their most recent data suggests that the Quantum Kinetic Well may perform sequestration of CO_2, removal of CO_3, pH enhancements, water purification, and biological stimulation in the form of protein-matter growth, humic material, and phytoplankton fluorescence. Furthermore, the device may reduce radioactive elements such as tritium, strontium, and other noxious elements in the seawater. It even appears the device may stimulate a full carbon-cycle within the treated dielectric, something no one thought possible.

* * *

With a whirlwind of momentum behind him, McKane applies for the X-PRIZE Carbon Removal project sponsored by the Musk Foundation. This application marks McKane's first opportunity to *teach the tech* outside the confines of his limited social media platform. McKane understands there is little, if any, chance his application will win, or even be acknowledged by the team of reviewers. There are a thousand other applicants. Most are better funded, politically connected, and far more professionally resourced.

After all, McKane is not even a trained scientific writer. He speaks with the voice of an artist, not a nuclear engineer. What foundation board would even pay attention to his propositions?

Still, McKane has secured every previous victory on his journey only through stubborn, patient persistence. He will no longer refuse to explore every avenue available to him, or let disappointment deter his ambitions. McKane determines to use the X-PRIZE application process as an opportunity to spread the word—one company, one patent, one person at a time.

His submission is intriguing and voluminous. Backed up by research from a national laboratory and a prestigious university, McKane assesses that the goal of carbon removal is possible by relying on the resonant quantum transformer to recalibrate nature's own carbon-cycle. By strategically placing commercial sized units of the Quantum Kinetic Fusor and the Quantum Kinetic Well where the largest rivers meet the oceans, biological stimulation would progressively normalize marine CO_2 levels through a recycling process.

The approach would function as an atmospheric time machine, in effect taking the environment back in time to when it was pure, clean, and supportive of all life. The synthetic machine would also chemically increase the ocean's capacity to absorb carbon from the atmosphere and safely store it within its depths, where the life cycle would begin again, a world reborn.

15

Hollywood Calling

Before McKane became the inventor, he was an avid consumer of science fiction literature and film. His favorite fictional character was *Iron Man*, Tony Stark, first introduced to the world by Marvel in 1963. Comic book lore depicted Tony's father as a notable inventor who, inspired by the idea and hope of sustainable fusion power, developed an electro-magnetic machine that could provide limitless clean renewable energy to the world. Although the project ultimately failed to deliver the promise of fusion power on a large scale, its underlying technology was used by Tony Stark to build a modular version of a fusion power plant–the Arc Reactor, the source of Iron Man's secret powers.

Over 50 years of superhero comic book tales and blockbuster movies, the saga of Iron Man's Arc Reactor has inspired many real-life inventors to follow the same dream of fusion power for the betterment of mankind. Recently, MIT has been working on a reactor that would function like the arc reactor, although the science team avoids using the same name.

***

McKane can't advance the adoption of his working version of the Arc Reactor without a comprehensive plan to *teach the tech.*

As a child, he learned in Sunday School that there was an original Great Commission, time stamped 33 A.D. This brief dictate of the Christian messiah embodied a resurgence of hope after an apparent tragic ending to a promising story of universal redemption.

Two thousand years later, the world seems locked in a state of perpetual conflict over dwindling energy resources. Perplexed by the dangerous state of international politics, McKane wonders if another critical commission is needed. The first Great Commission was propelled by word of mouth across ancient empires to the ends of the earth.

The second must spread more quickly, with a sensitivity to the impending environmental crisis facing humanity.

McKane resolves to work with anyone willing to accept the data without muffling inquiry based on their first impression, or the biases of prevailing scientific dogma. Advancing the technology will involve a cast of supporting characters, a massive funding effort, wonderful machines previously dreamed of only in science fiction, and episodes of joyous triumph along the way.

But McKane has no idea what this yet-to-be-assembled team of scientists, spokespeople, and investors will look like. To temper his expectations of positive reception from audiences across the country, McKane continually reminds himself of the implausible, *avant-garde* nature of the work he's representing.

Nevertheless, he hopes for an infusion of new energy and perspective into his work from a supporting cast of enthusiastic partners.

* * *

One day, he is meditating after a long day in the laboratory. It is a practice learned in the Happy chapter of his life—before the accident.

McKane thinks of meditation now as a non-traditional therapeutic practice inoculating him from long term PTSD. It provides him stress reduction and timeless moments of mindfulness, but rarely offers him fresh insight or revelations.

But today is different. While meditating, McKane sees an image of a face and the sound of a name within a whisper. The vision is of his old friend Greg from ProVault NW.

McKane knows that he must reach out to Greg.

* * *

McKane messages Greg in the morning. They reminisce by phone about their good times at the vault club. Eventually, the conversation turns to personal updates and current events in their lives.

McKane asks about his friend's situation after college. Greg reports working with Russell Bobbitt, the actor and legendary prop master for Marvel and Disney. Russell's movies include *Guardians of Galaxy, The Avengers, Iron Man, Thor*, and many others futuristic thrillers. Russell designs and builds original movie props by merging fantasy with reality. His products capture the public's imagination about science, fictional heroes, battles with exotic weapons and armors, and imaginary spacecraft powered by fusion energy.

The Arc Reactor used in the *Iron Man* movies is one of his fantastic creations.

Stunned by coincidence, McKane describes his recently patented invention that generates a magnetic field while ionizing the ambient air at 12v input, like the theory behind the fictional Arc Reactor. Excited, Greg promises to inform Russell of McKane's work.

The Arc Reactor sustaining plasma fusion (Bremsstrahlung Radiation)

It seems days, but it is only hours, until Greg reports back. Russell wants to talk to McKane and soon.

For a moment, McKane feels some concern, a twinge of paranoia. He considers the risks of a meeting. *Will these massive companies sense I am infringing on their intellectual property rights? Is Marvel about to marshal an immediate "cease and desist" court order on the new technology? Do I need to call my patent lawyer?*

Sensing a need for caution, McKane asks for more context from Greg before agreeing to a meeting. He inquiries about the reason behind Russell's request to meet, confident Greg will tell him straight.

"When Russell learned about your patent, he seemed staggered," Greg says. "He just wants to talk with you to find out how it works."

McKane considers the potential advantages. *What better way to communicate to the world than through large international corporations capable of the mass production of books, TV shows, press conferences and blockbuster movies?*

"Let Russell know I would be delighted to meet with him at his convenience," McKane says, taking a first confident step into an entrepreneurial world of communication by film and storytelling.

Russell and McKane meet by Zoom call. Russell's warmth and wit immediately reassure McKane that this will be a friendly meeting, focused on the new technology and its applications rather than any concern over intellectual property rights. Russell is not interested in such IP matters, in part because no one has patent rights on a functioning Arc Reactor.

No one, of course, except McKane Lee.

Russell explains the reason for the meeting.

"I see what you have done with your technology and the Arc Reactor – very smart. I love your story. I was excited to learn someone finally figured out how to make a real Arc Reactor. I have been waiting for years! I knew it was possible. When I heard about your patent, I called my father-in-law, Michael. He is a recently retired executive with the USPTO. And a member of the Intellectual Property Hall of Fame at the federal agency. Michael was surprised to hear news about your invention. He revealed few inventors are awarded a patent in just five months, especially for a new kind of technology. He concluded "somebody must be looking out for these guys."

"Interesting. What can you tell me about your involvement with the Iron Man story and the Arc Reactor?" McKane asks.

"The movie version Arc Reactor was designed and built in my shop after talking with scientific consultants and brainstorming with Robert Downey, Jr. The ideas behind the reactor were electron capture, electromagnets, and fusion."

"Some very special electromagnets, I assume."

"Exactly. The scientific premise is controlled fusion energy as the source of Iron Man's secret power. It is also the only way to sustain Tony Stark's life after he incurred a serious injury in warfare. The device controls his every breath and protects him from embedded shrapnel, lodged too near his heart."

"I remember the part about the device pulling the shrapnel away from Tony Stark's heart. I will have to work on that," McKane says, smiling.

Russell further explains his work. "It all starts with a scientific proposition, even if it is never developed into real technology. The public bought the idea of the Arc Reactor, literally. The toy version was mass-produced for sale. Disney's target audience has always been the 7-year-old child. Children believe in the possible. Every child who buys a toy Arc Reactor is affirming the possibility of a future that includes it."

Russell turns the conversation to McKane's journey.

"I hear you locked yourself up in a garage for five years and worked on your project."

"That's all true."

"I am so envious. I wish I could do that. I am an inventor, too. I dream of escaping into the solitude of my laboratory for five years and design an all-electric car that travels above the ground. Maybe I'll do that when I retire, not too long from now," Russell says.

Russell asks about specifics. "What about the energy output of your machine? Is it capable of operating a large manufacturing factory or a small electrical power plant?"

"In the laboratory we work with a modular, tabletop unit. To run a small power plant, you would increase the size of the electrodes and wiring configurations to extract more electrons from the water, thus creating more electrical current. The extent of energy harvesting is only limited to the size of the electrodes. This means our quantum system can also be engineered to run large power plants."

"How much does it cost to convert over to your electric system?"

"It depends on the existing business infrastructure. Many are ready-made for it now; so, it's plug-and-play. Some will require more redesign work. Ours is a retrofit technology for use in any energy demand situation, including internal combustion engines, rocket propulsion thrusters and regional power grids."

"Could it be used as a car battery recharging station?"

"With our prototype we generate amazing amounts of energy at only 1.9 watts. Other versions could be designed to power batteries of all shapes and sizes."

Russell cannot hide his enthusiasm for McKane's work.

"What you are doing is so much bigger than science fiction and selling toys. It is even bigger than a working version of the Arc Reactor. It is a system to power the world, a new form of electricity. This will be scary to some, exciting to others. Now you need to market your invention.

Think about how Elon Musk became successful. I was in the room with him when he first said his rocket was flying to Mars in 10 days. Of course, that came true. Elon explained that when he was a young man, he pledged to never set off a rocket smaller than the last. He has been true to his promise. Elon has promoted his business through smart applications of cutting-edge science coupled with the power of imagination. You are in a similar position, except for the whole gazillionaire part. Your invention can change the plane of history. It seems you are the z-axis of electricity."

McKane feels as though he has landed a crucial and understanding supporter of his work in Russell Bobbitt.

Russell affirms McKane's impression. "If you want our help, we are all about that."

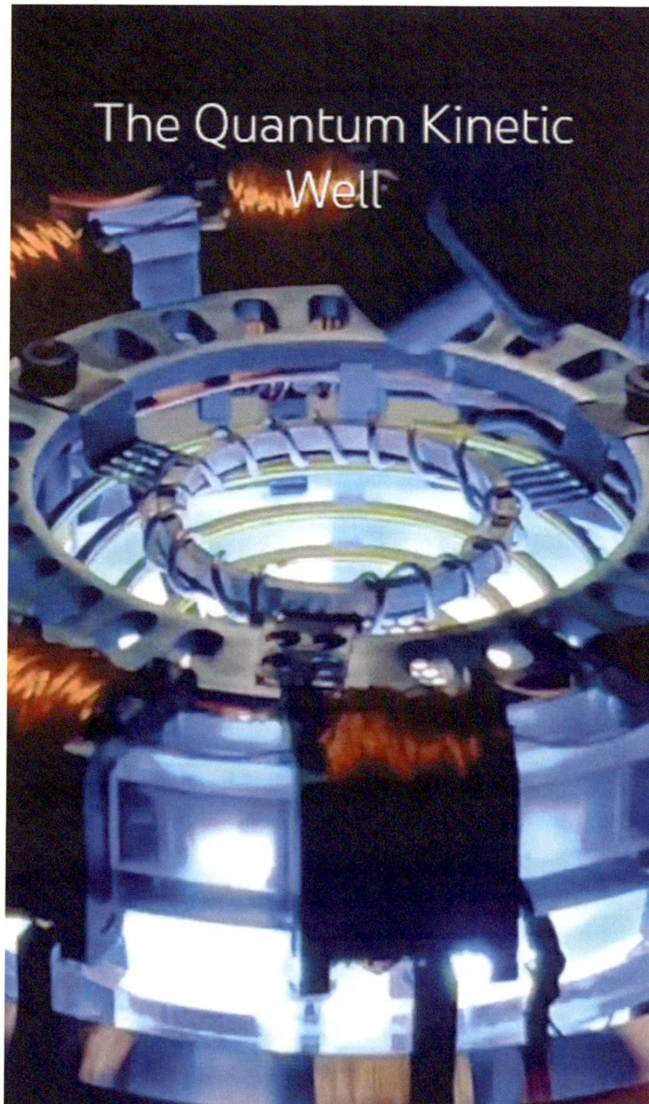

The ARC REACTOR device seen on Quantum Kinetics' YouTube Channel

16

On the Z-Axis

After his Zoom call with Russell, McKane is propelled into fresh reflections about his work. He understands Russell to be making bold predictions about an emerging technology.

McKane lets himself wonder.

Should this device really be considered a new form of electricity? Does anyone really merit being called 'the Z-Axis of Electricity'—let alone a newcomer to the world of scientific discovery?

McKane understands, along with all first-year geometry students, that Cartesian Coordinates in three-dimensions allow for ordered triples of real numbers to be displayed within three perpendicular planes intersecting at one point, the origin (0,0,0). This three-dimensional space, created by merging a z-axis into a simple x-y dimensional plane, allows for a vastly more complex and varied visualization of possible locations.

Russell's reference to the "z-axis of electricity" staggers him with a newly considered possibility.

Is it possible the Quantum Kinetic Well operates at a location along an infinite axis in space never identified, where energy access is as elementary as engaging a resonant electronic switch, just as predicted by Maxwell's thought experiment?

Stunning realization marks another inflection point on his journey. McKane senses the practical effects of his next decisions will surely be momentous. He is sure about only one thing: the vast capabilities of the tiny machine existing on a table in his laboratory.

Under the influence of newfound support from a voice in Hollywood, McKane is bursting at the seams to take the technology into development and marketing. Still, he does not want to make the mistake of the novice traveler in an unfamiliar wilderness, a hasty decision leading to sudden catastrophe.

So, he turns to someone more familiar with the landscape of technology development and commercialization, messaging his friend Matt. Luckily, the entrepreneur is available for another Zoom meeting the following week. They agree it is time to discuss recent events and the test results at the national laboratory.

McKane is always excited to talk with Matt, who first granted him audience only as a favor to Dr. Pollack. Matt's involvement with McKane has since been as a professor to a protégé. Matt offers advice not as an official representative of the giant corporation; rather, it is his personal interest project.

Their conference is scheduled for an hour but will likely take longer given what is at stake.

"It looks like you have been rubbing shoulders with some very important people—Mr. Bobbitt and Mr. Kirk, impressive. It would be more impressive if your goal was writing science fiction," Matt says.

McKane picks up on the sarcasm.

Matt continues: "You need to figure out the minimum moves to checkmate. The challenge when starting something brand new is convincing the scientific world what you have. You need to continue with your foundation based on simple, scientific validation. You have already taken on the research with PNNL. You have the data. You are getting more data to fortify your research. Your evidence is repeatable. You have two patents. Now is not the time to be swayed by the lure of glamorous people and places. You first need to do something unequivocal in physics."

McKane is listening to every word and knows not to interrupt priceless advice.

"It is not yet obvious what you have done. Tell the scientific world up front the most shocking thing about your discovery. Publish the data in a major scientific journal. Aim high for the most important journals. Why not *Nature* or *Science*? Both are audacious goals. Submit the article and ask for a meeting with the editors. Explain how it will bring prestige to their journal, not depreciate them. If you can land even one of your transmutation claims in your article, that would be big, a career-making event."

Matt adds a warning. "It will be very hard to get the scientists on board. They will criticize your claims. Worse, no one will believe you—especially the claim your device makes excess energy. That is not accepted science. They may not believe you even if it is published."

As if in answer to an unstated question, Matt counsels McKane on how to deal with rejection from the scientific establishment.

"You have a very big idea, but don't force it down anyone's throat. Be the humble inventor. Simply explain this is what you did, and this is what you saw. Be patient with them. Understand they are very smart and well-meaning. When you are talking about your transmutation device, don't preach any scientific theory underpinning the process. Your foundation is data, not concepts. Be careful not to say this device is the "only way" for transmutation to occur because that will offend the sensibilities of the physicists and astrophysicists."

Matt makes a prediction. "If you can get the scientific establishment behind you, you can change the world. If they believe it is really transmutation, then they will not slow walk it. Ultimately, they will follow the best technical talent. They will like your affiliations with PNNL and UW. It is better to have the patents than not. You already have a massive head start."

Matt breaks from his presentation of counsel to McKane to ask him a question related to cold fusion. "Did you know that the Sun cannot produce nickel? So, where do we get it? It likely comes from the fusion power of a supernova. The data indicates your new machine rearranges atomic particles through transmutation. It supposedly forms new elements—including nickel—from basic well water. Do you claim to have a supernova in the chamber of your quantum well?"

McKane answers Matt without hesitation, as if he had pondered the question a hundred times.

"Have you ever considered the power of a supernova can occur at the level of a single electron?"

There is a pause. For a moment, it seems Matt is unable to speak. Then the words come out. "McKane, your brain is way past the Matrix. We are all just trying to catch up."

A fiery universe of possibilities is spinning around McKane. He senses a planet extending its reach to the stars.

17

Ambush by the Sea

True to fashion, McKane is first to arrive to the meeting.

He takes in the storied establishment where the investor dinner has been scheduled. A century earlier, the pier supporting the restaurant's structure housed food processing and cannery operations that supplied seafood delicacies to international markets. This venue helped establish Washington State as the salmon fishing capital of the world.

Now the establishment offers fine dining augmented with a prodigious variety of fresh seafood specialties from local waters.

The hostess guides him to a table with an expansive view of the Puget Sound. He thanks her with a smile, mindful to mention her name at least twice during the brief conversation.

That's everyone's favorite sound. Echoes of their own name.

Settling into his seat, McKane notes that it is an unseasonably warm evening for early spring in the Pacific Northwest—perhaps a daily record. It somehow feels even warmer at the table.

McKane removes his dinner jacket, hoping to lighten his mood. Tonight, he wants to set a cool and casual first impression. He's expecting an important guest.

His window seat faces due west with a view opening out onto the sound. The ferry traffic shuttling between Seattle and Bainbridge Island punctuates the quietude of the evening. In the distance, a bold outline of the Island emerges from carbon blue waters. Golden light from the setting sun frames the dramatic ocean-scape. Below the restaurant, boisterous waves dance against massive pilings at hightide, loudly clapping as if in celebration of winter's end.

Though early, McKane hopes there will be time enough to brace himself for what's about to transpire.

The executive partner of a massive venture capital firm has beckoned him. It will be their first in-person meeting. For months, the Venture Capitalist (the "VC"), whose investment group has nearly a billion dollars to spend on emerging energy sector companies, has been inquiring about McKane's newly patented technology. The VC has hinted at his firm's interest in exploring a formal relationship and has asked for technical data backing up the patents with USPTO and CIPO, including any recent experimental results supporting the transmutation research.

McKane ran an internet search of the VC firm's background and reputation. It has financial roots stemming back to the early 1900s, its wealth based on generations of successful entrepreneurs. Few people would recognize the name of the firm, now a clever acronym. But most would recognize its founders' names, ranking among the richest and most influential dynasties in American history. Now a global company, the present executives troll the world for viable investments.

But behind the firm's storied history lay risks to the entrepreneurs that land on the firm's golden brick road. Rumor has it that once the giant firm acquires a company, their initial concern for the ongoing enterprise and its founders ebbs quickly. And when the acquired company's initial performance drops below par, corporate liquidations soon follow.

Once inside a room, an 800-pound gorilla like this VC firm can stomp you.

McKane calls the waiter over, intending to order a glass of wine. Instead, he finds himself ordering a Jack & Coke with a twist of lime.

McKane has already provided the information requested by the venture capital firm. In fact, he turned the data over without hesitation. He now realizes he should know better.

A wise businessman would slow down, keeping trade secrets close to the vest until the strategic need arises. But McKane is an artist and a scientist—not used to a world of financial sharks. An enthusiastic collaborator and learner, McKane's posture has always been openness and honesty about the technology. Indeed, he is eager to share his novel insights with the world, even more so since adopting a mission to *teach the world*.

If the VC or anyone else wants an explanation on how the magnificent apparatus works, I can do that. Just give me fifteen minutes, he thinks.

Now McKane wonders, less confidently, if the venture capitalist has a new message for him or may even be personally bringing a "term sheet," the first legal document establishing conditions of association between investors and an entrepreneur. What the actual message or purpose of their meeting is, the VC won't say.

McKane only understands that the VC's message is important enough for him to fly in from California.

McKane's guest is twenty minutes late. The inventor wonders if the walk from the downtown hotel room to the waterfront is taking longer than expected. Perhaps his guest is lost along the maze of waterfront buildings.

The delay affords McKane time to think retrospectively about the project. He believes wholeheartedly in the new technology and has held that conviction since the day of its discovery. The present problem is financial. Without an infusion of new capital, there is a threshold beyond which he cannot pass. He has already spent more than a quarter million dollars on institutional research and legal fees establishing intellectual property rights. His savings are nearly exhausted. Funds seemed adequate at the beginning. Now they fall well short of upcoming expenses to advance his work further.

The head teller at the bank has expressed concerns about McKane's ever-dwindling bank account. He tells her honestly about the research and how it may help the world someday. He jokes about the monthly wire transfers to the Department of Energy, suggesting he is doing his part to fund the US Treasury.

The VC finally walks into the restaurant—thirty minutes late.

McKane stands to welcome him. They shake hands. The VC's grip is ferocious. As McKane's guest takes a seat at the table, he apologizes half-heartedly for his tardy appearance.

The waitress returns to serve McKane his drink, just in time. He takes a sip, and then another.

The VC orders his favorite Martini. To McKane it sounds complicated, like a medicinal formula or the "Vesper" version of the Martini from "Casino Royale."

He can't stop himself from imagining that the man across the table is on a secret mission. His guest has brought along a slim black briefcase. Could documents pledging the needed financial commitments be inside?

The VC requests the dinner menu, noting to the waiter that he is only interested in the fresh seafood. Without conferring with his dinner companion, the man orders hors d'oeuvres, a full platter of fresh oysters from Willapa Bay.

McKane picks up on his guest's vibe. The man's tone with the server is laying the groundwork for a take-no-prisoners conversation with the inventor. It strikes McKane that beyond the first impression, he knows next to nothing about the VC.

Conspiracy theories flash through his head.

Could this man be a corporate spy? Is he a covert agent of the fossil fuel industry aiming to undermine the game-changing technology?

McKane lets his rationality prevail. These dark theories have no underpinning evidence.

The VC jumps right into the evening's business, pushing for updates on the quantum research.

McKane dutifully provides the details, touting two additional patent applications in the works, more stunning discoveries in the national laboratory and progress with a journal publication.

Their server delivers the local oysters on an ornate silver platter just as McKane finishes his opening monologue. McKane's guest surveys the food.

"My research article was accepted for review by *NPJ Nature Magazine* in just five days," McKane says.

"I don't believe you. That's impossible. It never happens like that."

"I have the notice right here on my cell phone. You can see it if you like." McKane slides his phone across the table.

The VC steals a glance. "I still don't believe you. It must be some mistake."

McKane registers the insult, a glancing blow.

]"Even if true, I doubt the assigned reviewer is even a mainstream scientist," says the VC.

McKane senses a cold, harsh slap to the face. "To the contrary, the reviewer is a worldwide expert. He has been cited in scientific publications over 350 times. Two other reviewers have also been assigned, both with impressive scientific backgrounds. All three scientists express great enthusiasm about the research."

The VC looks unimpressed. Turning away from his dinner companion, the man orders a glass of Washington white wine.

As the server swifts away the empty silver platter, the VC says, "I read the draft of your journal article. It is pure gibberish. I don't understand a word of it."

McKane swallows hard. "Is that your opinion, or do you base it on the review of your consulting PhD physicist?"

"No, I didn't have her read it."

The VC signals no embarrassment for his lack of due diligence. "It doesn't matter, I don't see that you have a product meriting investment or even a scientific review."

Before McKane can respond, the VC interrupts their conversation to order the fresh Halibut. The server stares at McKane.

Unnerved, McKane glances absent-mindedly at the menu. He orders the first item that meets his glance, the Shrimp Tempura.

He tries once again to make some headway with the VC.

"Would you agree that transmutation itself is a game-changing process with applications across countless industries, easily worth billions or even trillions?"

The VC pauses and takes his time to order another glass of white wine.

"That's interesting, but I still don't believe you."

The discussion is going nowhere fast. McKane tries to regroup his thoughts before this potentially promising meeting, and his chance at landing a powerful investor, goes down in flames.

The VC is now sampling an Oregon Merlot. McKane speaks again, this time in a more confident and direct tone.

"We are scheduling another research project over in Richland, at the nuclear facilities. We have already talked with three of the lead scientists there. The suggestion was made that our device might offer a secondary water treatment system to mitigate the release of radioactive wastewater into the Columbia River. If so, it could address impacts from nuclear catastrophes, such as at Fukushima. We also think it may transmute water into tritium on demand. Tritium is worth $30,000 per ounce. It is one of the biggest line items on the Department of Defense's annual budget. While testing, we plan to carefully monitor the extent of excess energy and hydrogen produced by our device. Basically, our apparatus is a pristine energy generator without residual pollution."

The VC sits silently for a few moments. He seems distracted by the first bite of a deliciously prepared halibut. McKane notes how long it takes him to savor the first bites, before finishing with a long sip of Washington white.

The VC looks squarely at McKane, as though prepared to end a battle of wills with one swift stroke.

"Don't go over there to the nuclear facility. Don't do any more testing with the government. They can't help you. It's not the right time."

McKane notices there is no offer of proof as to the meaning of *"the right time."* An internal alarm in his mind is activated. He's staving off a rush of panic in his brain. *The contest is not over*. His chest tightens. He prepares for the next blow.

The VC delivers it without blinking. “You are not even a scientist. You have no credentials to make your claims or go forward with more research.”

McKane sits back in his chair. He could deliver a clever comeback, but instead he deflects with a simple admission. “That’s true. I *am* an artist turned inventor. I just happened to make an important scientific discovery.”

The VC leans forward. “So, why should anyone believe you? You claim the ability to dissociate any dielectric with your process, transmuting it into new elements. That’s impossible. It’s a *hoax*. Whatever you are doing, I challenge you to dissociate D2O, the standard dielectric used in cold fusion experiments. No one can dissociate heavy water.”

McKane leans in, too.

“I can,” he says firmly.

“I don’t believe you.”

“Give me a week, and I will *show* you.”

Backed into a corner, the VC ignores the offer. Instead, he blurts, “What I really want is for you to explain to me how you do transmutation. Give me a simple formula, *A+B=C*. Tell me in a few words and put it down on three-or-four pages max. Send it to me. Then maybe we can talk further about your future.”

McKane recognizes the ambush almost too late.

Of course, it’s a trap. It only makes sense. Transmutation and cold fusion will be a 40-trillion-dollar industry. There are bound to be competitors out there searching to exploit an opportunity, especially from an inexperienced businessman.

McKane leans back in his seat again. His eyes fall to the nearly untouched plate of food in front of him. He wishes he could be somewhere else, drifting along the z-axis in deep space in his own flying saucer.

Wanting to bring the engagement to a swift end, he contrives a hasty excuse about the lateness of the hour. He stands and thanks the VC for the opportunity. Wishing safe travels to his dinner companion, he exits the restaurant, leaving behind a meal uneaten.

There will be no rematch.

* * *

The sea air is pushing in heavily from the northwest convergence zone, bringing with it cool relief from the heat in the restaurant. McKane's steps to his car are swift and buoyant.

He notes his appetite has suddenly revived.

18

Vindicated Vision

A lesson learned.

In the wake of the disastrous meeting with the VC, McKane now sees the investment space he's entering for what it really is—full of great opportunities, but also threats. He'll come across potential allies. As well as charlatans.

His most recent guest had come to Seattle under false pretenses, disguised as an inquisitive investor. Perhaps the man was a corporate raider, dangling massive resources just out of reach to entice disclosure of closely held trade secrets and revolutionary experiments.

McKane muses that *VC* could stand for "*Vulture Capital*" just as easily, the kind of firm that circles at a distance until ready to pick up the remains.

It's time to get back to work.

* * *

The first phase of McKane's business plan is complete. Now, McKane must set himself to a crucial task on the trajectory toward success, that of reviewing and refining the journal manuscript submitted to *NPJ Nature Clean Water*.

The renowned publication accepted his scientific article within five days, an amazing accomplishment, and a great sign for the novice inventor. The article has now been assigned to three esteemed scientists for review and editing, each having different technical backgrounds. McKane wonders what the swiftness of the process portends. The future of his work now rests in the hands of this select group of scientists. For McKane to make headway at bringing his invention to market, brilliant scientists must formally recognize the merits of the discovery.

He sits and waits, secreted away in his scientific monastery. While he waits, McKane experiments with the acceleration cycle of his retrofitted, hybrid, hydrogen-fueled BMW engine. He comes to think of the novel engine as powered by the energy of the Sun.

With the balance of his time, he designs another quantum energy experiment, this one even more radical than the last. He predicts the newest experiment will yield the most shocking revelation of all.

* * *

An email notification flashes on his phone. The reviewers' comments from *NPJ Nature* are coming back early.

He hesitates momentarily before opening the message, realizing the significance of this moment. He may be crossing another consequential threshold in his inventor's journey. McKane wonders if the feedback from this publication will further shape or clarify his destiny.

Opening the formal messages, he reads the first round of "Comments to the Author."

Reviewer #1

The manuscript on A Novel Resonant Electro-Physical Transmutation Method for Water Purification, by Lee et al., is a unique paper with two patents already published online on water purification technique. The indigenous tri-coil design was designed to overcome the challenge in maintaining the resonance with the dielectric medium during operations. It is extrapolated that it will help scientists to work on a number of exciting new arenas of research where hydrogen production could be of help as well as radioactive tritium could be harvested as discussed by the authors. The paper is well thought out with clear objectives integrating logics with experimental proof lending scientific credence to the manuscript. It is an original piece of work undertaken by the authors. The experiments undertaken for analyzing data reach to a reliable conclusion however the chemistry of the processes and their logical discussion is missing.

Reviewer #2:

The manuscript is technically and scientifically sound with alternative intensive studies on water purification using electrochemical methods that is purely physical and devoid of current passage through an electrolyte in the conventional electrolytic method. Average metal concentration of elements pre and post experiment values were determined via ICP-OES. The study had shown that utilization of the resonant magnetic fields in conjunction with electrostatic temporal flux lines aid in dissociation of both FW and SW, with no electrolyte at a very low current density (0.142 mA cm-2) as compared to prior technologies the water purification sector. It also favors the easy removal of toxic metals such as arsenic from fresh water. Going by the convincing results and discussion presented in this manuscript, I recommend its acceptance after attending to the minor corrections.

Reviewer #3:

The authors are commended for the outstanding research presented in the manuscript. The discussions of the findings of this study present a novel means to purify water; a much-needed invention throughout the world. The manuscript is well written however; the authors are requested to read it once more after amending the figures and table indicated above.

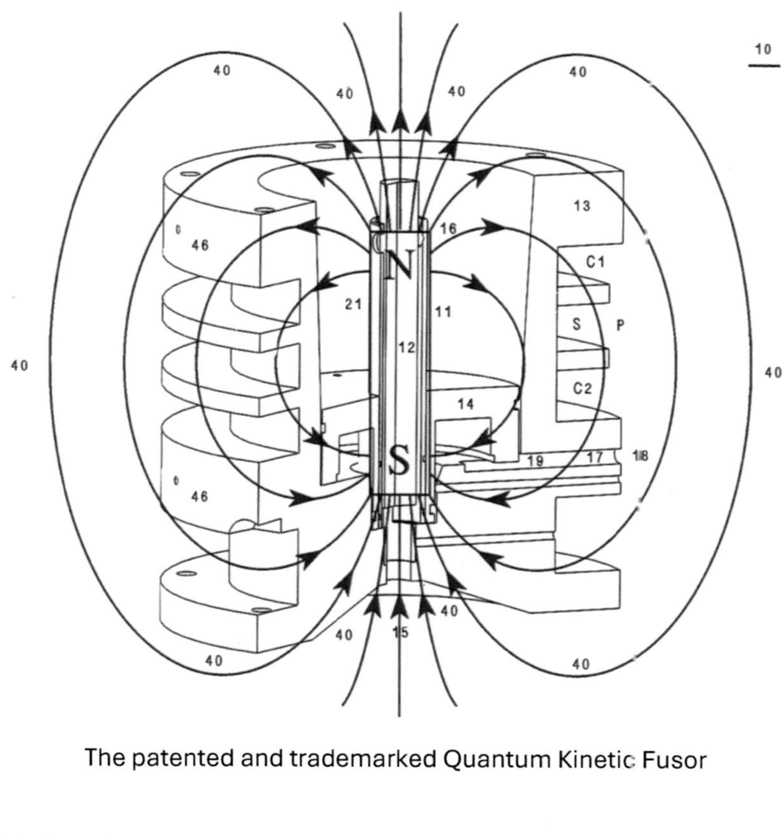

The patented and trademarked Quantum Kinetic Fusor

McKane is stunned.

He never expected this degree of glowing feedback from all three worldwide experts. He finds himself intrigued by their curiosity on the underlying process of the Quantum Kinetic Fusor.

They ask him for more details, please.

McKane has been waiting for this invitation for six years.

19

Star in a Jar

Going forward, McKane faces a simple but revolutionary objective: the need to establish transmutation, or "synthetic nucleosynthesis," in a way that can't be refuted.

If I can get scientists to understand the reality of nuclear reactions occurring at super low energy inputs in the apparatus, McKane thinks, *then my ideas could change the world.*

And he is convinced that he knows how it can be done.

He must run his experiments again, this time using heavy water (D2O) as the dielectric. He will follow a paradigm established by prior tests that attempted—but failed—to produce cold fusion. In the case of the quantum kinetic well, McKane is confident his results will be different, clearly distinguishing his machines from historical studies.

Interested but skeptical scientists will require him to show how the QKW succeeds in producing repeatable nuclear fusion, where all previous research projects have failed to generate reliable, useful results. If he can demonstrate how the synthetic device can overcome the impossibility of the Coulomb barrier—mother nature's repulsive force field preventing fusion at the atomic level—there's virtually nothing left to stop countless applications of the technology to better the world. `First, though, he must teach the world nuclear fusion as if it were first-year algebra.

$A+B=C$.

Despite setbacks in recent months, McKane feels mentally prepared for the challenge. In fact, he can hardly contain his excitement in anticipation of launching the new experiments. In the planning phase, the experimental design for the new trials somehow comes to his mind instantly, as if he had been planning it for years. For once, he senses the momentum of a resonant cresting wave of success just ahead.

That wave is about to break in a secret lab, steps away from the Pacific Ocean.

* * *

The experiment will be done at various energy inputs: 30 volts, 60 volts, 90 volts. Within each phase, run time will be 24 hours in the resonant cavities. With the staggered voltage inputs, the results should reveal whether the QKW effect is linear or exponential. He will use deuterium enriched heavy water this time, extremely expensive but containing double the neutrons of seawater or well water.

McKane wonders in retrospect why he used seawater, well water, and distilled water in all prior experiments. Perhaps he was over-confident in the machine. On the other hand, how magnificent would the outcome be if the new technology worked with basic, cost-effective materials.

Not everyone could afford to spend $5,000 per sample on deuterated water. But anyone could capture water from a domestic well or the seashore.

In McKane's judgment, his prior work—deemed promising but incomplete by the reviewer's standards—was not a waste. It was a revelation. The original experiment established with low power inputs of 1.9 watts or even lower, anyone could reliably produce element transmutation, nuclear signatures, and excess energy.

The problem was that the original water samples, though described accurately by PNNL in pre-test procedures, may be viewed by unconvinced observers as having a level of innate contamination. After all, they will say, isn't ground water "dirty" in some respects, and seawater potentially "contaminated?" What this means is that critical scientific minds might later question the experimental design as well as the results, no matter how remarkable.

The remedy to this conundrum is to use industrially manufactured deuterium enriched heavy water to 99.9% purity. Under these conditions, potential attacks by others seeking to weaken his research findings would be cut off entirely.

* * *

He will conduct the experiments over three days, not long after his cynical guest threw down the gauntlet at their meeting in Seattle.

"I don't believe you."

Those words ring out in McKane's memory, fueling this phase of his work.

He carefully sets up the apparatus and arranges the sensors in place. Next, he positions the electrical connections, requiring only a few minutes. Activating the quantum machine is even more effortless, requiring only turn of a user-friendly electronic switch.

Although the initiation phase of the test is parsimonious, what follows is incomprehensible. Inside the resonant cavity a sudden, complex, quantum kinetic change in atomic isotopes of all selected samples will occur.

McKane now thinks of the process as *a star within a tiny jar—the power of the Sun at the size of an electron.*

* * *

The inventor has never attempted transmutation at only 30 volts. He understands the scientific community will largely dismiss the possibility of a nuclear event occurring at such super low energy input. *That's absurd nonsense,* they will say.

Nevertheless, based on hundreds of trial runs with seawater and well water, he is confident the result of the experiment will prove the same—perhaps far greater in effect—given the density of neutrons in the heavy water.

McKane monitors the machine's operations with a remote camera for the 24-hour periods included in each trial. Three additional instruments will provide measurements for the test. He must take the greatest care to substantiate and document his observations continuously, laboring more than anything else to track the energy levels coursing through the Quantum Well.

In these experiments he is attempting something never done in scientific history: D2O meeting the quantum kinetic well phenomenon. As he begins the trial, McKane again finds himself holding his breath. At the threshold of the unknown he stands again. This time around, he may be constructing a new reality with galactic implications.

But he's come to embrace the flush of excitement that meets him in these moments. To him it's another vaulter's high.

* * *

McKane closely monitors the experimental results. Initial observations are more than intriguing. They are stunning. He can hardly wait for confirmation from the National Laboratory.

* * *

McKane drives across the state to deliver the experimentally treated water and sediment to PNNL. The trek takes place on a Friday morning, immediately following the final trial. He decides to keep the anodes and cathodes for later testing by SEM by a respected laboratory in a small community north of Seattle.

* * *

Nick, the team leader on this test, is there to greet him in the parking lot exactly at 9:00 a.m. as planned. McKane was instructed not to enter the facility itself, since there had been no time to set up the procedures to grant him a security badge. Though unable to visit inside the federal lab, the game-changing tests on the materials from the QK Well can proceed.

McKane catches Nick up on the publication status of his work. Nick can't conceal his surprise that McKane's journal article came to be accepted for review by *NPJ Nature* so swiftly.

"I still can't understand why you got accepted by *NPJ Nature* for review in a matter of a few days. Most scientists that attempt to get published in *Nature* are never accepted over their entire careers. They don't even get to the review stage. Less than 15% get past the first hurdle. You are not a scientist nor an academic, but your article was immediately accepted for review. I have been trying to get published in *Nature* for 10 years and never got a look."

"Beginner's luck," McKane says, hoping to convey empathy to a fellow scientist.

The truth is, he feels both proud and daunted by his success so far.

"I'm keeping my fingers crossed, Nick."

"I will get right on the tests starting today," Nick advises. "We will first have to dilute your water samples one thousand percent, or your treated water will destroy all our instruments. Full analysis may take two months. Stay patient. This is going to be important."

* * *

After leaving PNNL, McKane is first in line at the ferry terminal in Kingston. He will take the next arriving ferry on the 30-minute ride to Edmonds.

As he waits, a police officer with a Homeland Security Badge approaches his vehicle. Her companion is a large police dog, a black Labrador mix. She motions for McKane to roll down the truck window.

"You're not in trouble, and you are not under arrest. I just must make sure my dog gets his treat for the day. Are you carrying a weapon or any ammunition this morning?"

"No ma'am," McKane says.

The Labrador is jumping against the truck's back door, attempting entry through solid steel. Frustrated, the powerful dog crawls under the truck's undercarriage, snorting and whining loudly.

A crowd of curious commuters looks on.

"Well, it seems my dog really likes your truck," says the officer. "Are you sure you don't have a gun or ammunition on board?"

"I don't even own a gun, never have."

With great effort, the officer drags the animal out from beneath McKane's truck. Immediately, the dog slams his head against the truck's back door once again, barking.

The officer appears nonplussed. She poses no further questions and apologizes for her canine companion. Then she moves down the line of vehicles waiting for the ferry.

The ferry arrives, a behemoth floating gently into the terminal stall. McKane eases his F-450 truck into position on the ferry ramp near the head of the line of cars. The ferry peacefully departs. Enjoying the commanding view of the water, McKane reflects on the Homeland Security dog's behavior.

Why would a trained police dog single out his truck as a potential concern for Homeland Security?

Then an idea forms in his mind.

McKane turns and scans the back of his cab. On the backseat are three sealed plastic bags. Inside the bags are the three sets of electrodes used in the recent transmutation experiment. On each electrode is a plating of some deep chocolate brown substance of unknown identity.

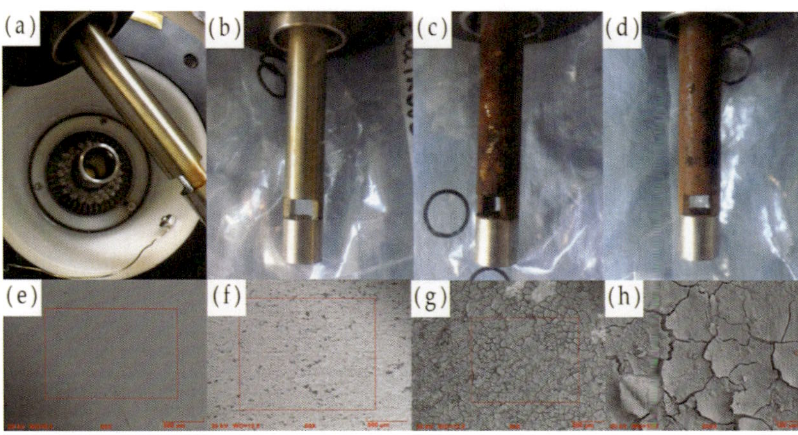

a.) D2O 30v wet b.) D2O 30v dry c.) D2O 60v dry d.) D2O 90v dry

What's in gunpowder?

McKane reaches for his phone and runs a *Google* search.

Moments later he finds that the general components of gunpowder are Sulphur, carbon (charcoal), potassium nitrate, and potentially thermite (the only known element that will burn through steel). All are by-products of prior experiments in the quantum kinetic well. Previous experiments consistently yielded Sulphur, chlorine, titanium, silver, iron, carbon.

Oh my god, the dog must have detected those elements, he thinks. *That Labrador deserves two treats for the day.*

McKane calls the ferry terminal to tell them. And to apologize.

20

The Test

McKane's hopes are high for the pending data release from the D2O experiment. Waiting isn't his strong suit, and patience isn't his virtue. He takes solace in Nick's reassuring words, *stay patient—this is going to be important.* Nick forecasts a timeframe of at least two months until the results come through.

***

With experiments in the lab at full stop, McKane has time to glance backwards. He reflects on his long-term obsession with nucleosynthesis, the natural nuclear reaction that binds a large atomic nucleus from two or more smaller ones. He remembers well the day its power first manifested in his make-shift laboratory. His initial shock at the instrument readings was soon overwhelmed by a penetrating question, "*Where is all this energy coming from*?"

Little did he understand at the time that he had stimulated a natural nuclear event within the confines of a six-by-six-inch chamber on his benchtop. After years of research, he now understands that the small synthetic device holds the promise of efficiently conquering the

Coulomb barrier and thus stimulating nuclear fusion with little electrical effort.

No one had figured it out before. For seventy years, scientists had thought the only way to defeat mother nature's repulsive force field was to smash atomic particles together at super speeds and/or million-degree temperatures. The Sun was their model. Afterall, the Sun is an immensely hot space body and gigantic nuclear fusion reactor. However, after a generation of colliding particles and mega-million-degree temperatures in sensational facilities, no one could sustain fusion ignition for more than a fraction of a second. All the billions of dollars spent in the effort didn't buy anything of commercial value.

Previously unknown to science was an alternate way to bridle nature's Coulomb barrier. Having his mind on gravity for so long, McKane understands this power to be the only innate process known to de-shield the Coulomb barricade. And when creation's atomic blockade falls, fusion rushes in bringing with it limitless plasma energy.

***

As predicted, the lab reports trickle in slowly over a two-month period, and even longer. The unfolding data presents an intriguing pattern of isotopic changes in the dielectric (and even within the electrodes)—a result never seen before in electrolysis-type tests—nor in prior cold fusion experiments.

McKane begins to expect the unexpected with each tranche of new data.

Nick apologizes for delays in finishing the analysis on schedule. Some of the treated D2O samples were analyzed twice. Nick explains the initial findings were so unexpected the National science team was shocked into a state of disbelief. They admitted to initially not believing their own instruments—the best and most capable in the world! And when the team ran the tests a second time, the results were the same! No one had ever seen this before.

***

Nick supervises analysis of the post-test samples and reports the data.

McKane composes the narrative summarizing the experimental design, tests, and findings. The paper describes an electron-altering process that stimulates micro-supernova star events like single electron Nucleosynthesis. The resulting physical conditions promote synthetic CNO burning, electron extraction, and room temperature fusion.

Both McKane and Nick understand the publication of the research paper will change the world of science—and human history forever.

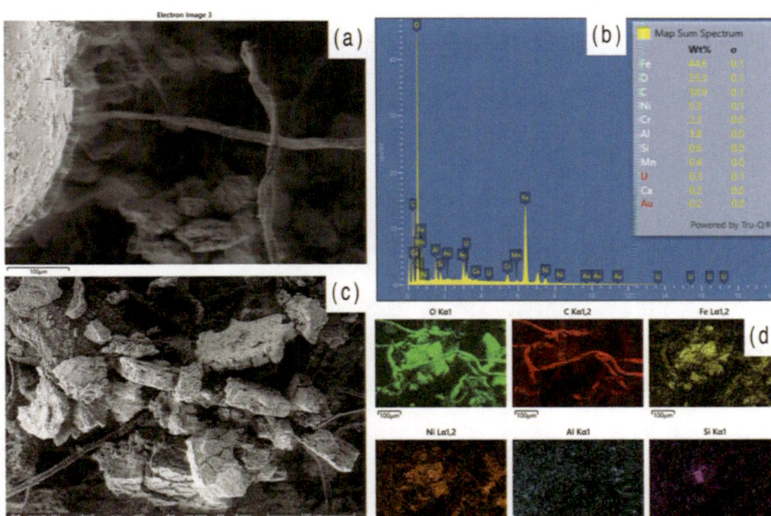

Supplemental Figure 7. Sediment scrubbed off electrodes. **(a)** SEM image of dried sediment. **(b)** EDS values. **(c)** SEM zoom factor 500x. **(d)** MAP readings of sediment. Nano-particles of Al, Si, C, O, Fe, Ni, Au, Ca, and U. The sediment is ferrous and is attracted to a neodymium magnet. Woven shaped Carbon like tentacles detected.

D2O sediment fabrication (Organic & Inorganic)

21

Censorship

New doorways are opening to McKane. Now that his patents and publications are available for everyone to see, he regularly receives unsolicited inquiries regarding his mysterious technology. Messages roll in from people curious about his inventive machine. *When will it be available?* they impatiently ask.

Other contacts come in the form of offers to explore commercial relationships with technology companies, world banks, and foreign countries. He is asked to write a press release for two major newspapers in the United States. Two documentary filmmakers reach out to him wanting to tell his story to the world. He is invited to be a presenter at an annual climate conference in Europe. A famed tech giant comes out of retirement offering to help him save the environment.

McKane begins to accept he is no longer a recluse living in dismal isolation. No longer an unorthodox artist conducting self-taught experiments the scientific community will never recognize.

* * *

He receives an unexpected message from Roger, his patent lawyer.

Congratulations! I am pleased to inform you that we have now received a Notice of Allowance for another patent application. The rest of the prosecution of the patent application will simply be a matter of carrying through with some administrative formalities on our part. I am so glad for you.

So far, I think we are winning everything. That is due in great measure to the innovative ideas of McKane.

Roger's message draws an immediate response from McKane.

To win a race, it helps to get off to a great start. You did that for me, Roger...and I am forever grateful.

P.S. An FBI agent flashed her badge at me several weeks ago. She wants to be friends and meet often. I also had an unsolicited visit from a very pleasant gentleman in a large white van marked "National Intelligence." He stated he wanted to change out my underground internet cable because of a hidden defect.

Seriously, Roger, I am not paranoid - yet!

Best – McKane.

* * *

With so many positive developments stacking up, McKane wonders when his beginner's luck will run out. He recalls warnings from friends about the treacherous territory he is venturing upon for the first time. They explained that as an unpracticed inventor seeking recognition from academia, he would quickly engender opposition.

Fierce opposition.

One of McKane's mentors made specific predictions about how roadblocks would begin to line up between his work and its acceptance by the scientific community. He suggested that academics would try to show him up for ignorance of advanced mathematics and engineering principles. They would further profile him as an over-confident schoolboy, a hack who never completed his studies in the proper major at university. Others would take offense at his unpolished writing style, mocking his whimsical forays into futuristic pie-in-the-sky applications.

These prophetic words of warning echo in McKane's mind.

What you are doing will challenge the stereotype.

Expect people to be incredulous and dismissive of you.

You won't be the first.

* * *

After six years of solo research in his remote laboratory, the headstrong inventor is ready to present a renewable, clean energy platform to the world—and for the first time in history—a functioning Arc Reactor. Some people have told him, including the creator of the movie version of the Arc Reactor, "You are the real Iron Man."

More importantly, his work has birthed clean energy machines like nothing scientists have seen before. One invention cleans contaminated water—even radioactive wastewater—from rivers and oceans. Other devices built by McKane propel water-fueled vehicles with his inspired injector system. But chief among his engineering marvels is the Quantum Kinetic Well (the "QKW™") and the Quantum Kinetic Fusor (the "QKF™"). The two machines coupled together provide a "scalable" working fusion machine.

An achievement previously unknown in Physics.

McKane's eccentric pursuits have led to an unprecedented accomplishment, one that captivates the interest of friends, family, and several key academics. But that is not enough for this scientific maverick. He aims to raise the bar even higher—to be published in a *Nature* journal twice in one year.

McKane takes his passion straight to the front door of scientific publication. He submits his newly completed manuscript, "A Resonant

Electro-Physical Transmutation Method for Synthetic Nucleosynthesis," to *Nature,* arguably the top scientific journal in the world.

Though a bold step, the inventor reassures himself by reasoning, *what better way to spread word of these novel methods and gain respectability than from the most vaunted, peer-reviewed publication?*

McKane does not have long to wait before it becomes crystal clear his paper does not stand a chance at the front door, nor even the back door. A week after submission, *Nature* sends out the rejection letter.

> *Dear Dr. Lee. Thank you for submitting your manuscript, "A Resonant Electro-Physical Transmutation Method for Synthetic Nucleosynthesis," for consideration. I regret that we are unable to publish it in Nature. As you may know, we decline a substantial proportion of manuscripts without sending them to referees, so that they may be sent elsewhere without delay. In such cases, even if referees were to certify the manuscript as technically correct, we do not believe that it represents a development of sufficient scientific impact to warrant publication in Nature. These editorial judgements are based on such considerations as the degree of advance provided, the breadth of potential interest to researchers and timeliness.*
>
> *In this case, we do not feel that your paper has matched our criteria for further consideration. More specifically, while we have no doubt that your ongoing explorations of the unusual physicochemical behaviors exhibited by your 'Quantum Kinetic Fusor' will stimulate others to think about their possible origins, I regret that we are unable to conclude that the paper offers the sort of compelling mechanistic*

insights that would warrant publication in Nature. We therefore feel that the paper would find a more suitable outlet in another journal.

Please be assured that this editorial decision does not represent a criticism of the quality of your work, nor are we questioning its value to others working in this area. We hope that you will rapidly receive a more favorable response elsewhere.

I am sorry that we cannot respond more positively on this occasion.

Sincerely,

Karl, Physical Sciences Editor, Nature

Swift dismissal strikes McKane like an unexpected punch to the solar plexus. It is even more staggering in view of how other nationally recognized physicists and engineers had previously welcomed the research and his strange new insights into the quantum world. McKane finds his work and writing summarily dismissed without editorial review or even referral to a referee.

McKane shrugs off what could be taken as an insult from a leading publisher. Instead, he thinks about his father's life-long advice to his son, *"Never, ever doubt yourself."* The inventor sends out his manuscript to *Nature Energy* within minutes of receiving the first censorship letter, hoping for a warmer reception.

One week later, he receives a non-acceptance letter from the Senior Editor of *Nature Energy* oddly echoing the same theme from *Nature*.

Regretfully, we cannot offer to publish it. Decisions of this kind are made by the editorial staff when it appears that papers, even

> *when technically correct, are unlikely to succeed in the competition for limited space.... We are unable to conclude that the paper provides the sort of significant advance in technological understanding that would be of immediate interest to our broad readership of researchers in the energy community.*

Undaunted, McKane presents the manuscript to *Nature Communications*. This time he is not surprised by the whirlwind brush-off. However, the Associate Editor kindly recommends contacting its sister journal, *Communications Chemistry*.

McKane follows the referral and sends the D2O article to *Nature Communications Chemistry*. His hopes are higher this time, based on the specific referral from the sympathetic reviewer. However, a hurried email soon arrives spurning the research with the same cut-and-paste language as before.

In a last-chance effort, McKane offers the manuscript to *Nature Micro-Gravity*. He anticipates their enthusiastic acceptance of his work since his machine uses quantum gravity waves to stimulate low temperature isotopic fusion. But, the repudiation slip comes with breath-taking speed—in only twenty-four hours!

McKane's profound disappointment at the last message is real but tempered by a new curiosity. He evaluates what just happened in the rapid wake of his submissions to these publications.

Five rejections by notable journals in less than five weeks. Perhaps it is a record—an infamous one. How did the unified messages of dismissal come about?

He tries to fit together the pieces.

How can seven years of grinding research, bold experiments, and dramatic findings be met with wholesale and immediate rejection by orthodox science? Is the current scientific infrastructure a take-it-or-leave-it proposition governed by gatekeepers with the power to dismiss revolutionary ideas and methods out of hand? More importantly, what are the realistic prospects for human evolution if leaders are unwilling to step forward onto uncertain ground?

Upon reflection, the pieces don't fit at all.

Outright rejection from the echelon of scientific society despite doing something unequivocal in Physics–demystifying the enigma of low temperature fusion!

A sadder thought enters his mind, a catastrophic paradox in view of the potential unleashed through his ground-breaking work.

He may be able to stimulate the release of unlimited clean energy from nature itself. *But the world will never benefit from the discovery.*

* * *

McKane returns to his normal working routine in the lab, hoping to shake off the major setbacks to advancing and publicizing his inventions. He resumes work on another electron bifurcating machine—this one with four wheels.

Despite the importance attached to this latest phase of scientific innovation, the inventor finds himself distracted by disappointment and lost in thought. He can't help but dedicate thought to his next move toward an eventual checkmate of his critics and detractors.

The work must be recognized. It's potential tapped by the scientific world before this journey ends. His breakthroughs in quantum physics are not just a fascination, the latest novelty for physicists and chemists. Deep inside McKane is the conviction that these inventions pave a way forward, a way to avoid energy crisis and shore up mankind against some of the weightiest concerns looming on a future horizon.

22

Touching the Unicorn

Judgment entered in the form of a cold stack of rejections.

An esteemed community of physicists and engineers finds that orthodox physics will not include a low-power atomic manipulating machine capable of converting isotopes into new elements. Further, they find it impossible that a low-voltage transmutation device can convert contaminating isotopes into inert elements.

Hard as it is to believe, for this informal jury of scientists, *seeing is not believing.*

And there can be no appeal—even if the jurists disregard the factual record provided by McKane in his write-up of the results.

Faced with the slamming door of censorship, McKane wonders if going forward means constant contention with an array of influential scientists lined up against him. For now, he fully understands that an organized, academic-based adversary has the power to trample him and quash his case for electro-physical transmutation.

McKane is at another crossroad, despite having found the pearl of great price.

In his favor are three unexpected messages from successful voices in the world of technology and science that just dealt a crippling blow to his confidence.

"Too bad for Nature. Their loss, your gain," states a national expert in the capitalization structures of emerging tech companies.

"The scientific peer review process drives me nuts. It is all a game. But don't worry. There is a better way—having a smarter technology than the other guys." opines the founder of a world-famous technology company.

Nick at PNNL, a close partner in McKane's research, sends a text message of condolence: *"Welcome to my world, McKane. I guess you're not a deity after all."*

* * *

McKane is amused by Nick's heady comment about attainment of demigod status. He sees the double meaning in the message: poking fun at his research partner, while offering an oblique compliment to cheer him up. After two years of diligent research together, McKane now sees Nick as much a friend as a colleague.

McKane soldiers on, stirred by these tokens of timely encouragement. He's embracing a new strategy, to meet directly with a renowned scientist who is skeptical about the technology.

* * *

On the morning he leaves for the peninsula, McKane is thinking about the last time he interacted with Randy. It has been years since their first meeting.

Then, a group of local scientists and educators had assembled to hear a young man present data about a new form of atomic energy. The meeting was Randy's idea. As a university professor for over 30 years, he often took interest in promoting students that showed promise or had specialized skills.

McKane had both in Randy's estimation.

Nevertheless, the meeting itself had been a disaster. Within minutes of beginning his presentation on fusion energy, the enclave of scientists ripped into McKane, insulting him with inane questions and disparaging remarks. When one of the attendees shouted out "bullshit," Randy mercifully intervened, dismissing the hecklers with a fiery look and a wave of his hand.

Since that calamitous day, McKane saw Randy as a heroic figure. A professor, lofty but easy to know. A forensic scientist, legendary for solving complex puzzles. A public servant, sought out by state and federal government for consultation. A family man, gentle yet strong. A friend, affable yet always dignified.

* * *

McKane has been on the road for an hour after disembarking the ferry. He is traveling a scenic route along the Puget Sound, the most popular passageway to the western boundaries of the Olympic Peninsula.

Outside, the sky is engulfed in white light from a luminous sunrise. The ocean's edge is carbon blue and boundless. Seabirds meriting art prints by Audubon populate the shoreline.

McKane's destination is an abandoned warehouse near a remote fishing village. The rickety old building had long been used for boat storage by a once thriving fishing fleet. But today, the monstrous relic stands as a dismal reminder of an industry plagued by misfortune and resource devastation.

As the Ford 450 rounds the final bend just before the marina, McKane sees the massive structure perched alone atop a barren headland. He eases into a vacant, unmarked parking area, covered with weeds and tall field grasses. On the building is a large sign: “Under New Management.”

Beyond that, there is no sign of life to be seen here.

The front door is locked, but McKane has the key. Randy entrusted it to him for ready access whenever needed. The professor purchased this property to serve as an adjunct workshop for his fabrication hobbies and as winter storage for his fishing boat. There was also the obvious future commercial opportunity to renovate the building into a rustic bed and breakfast.

The high bank location commands views of the Strait of Juan de Fuca and the majestic San Juan Islands to the north.

Since that rowdy meeting with the dismissive scientists, Randy had followed with interest his protégé’s struggles to gain public recognition for his inventive machines. Although the two of them communicated on a regular basis, Randy had never fully grasped the underlying basis of McKane’s ostentatious claim to produce nuclear reactions at only 1.9 watts of electrical input. The premise seemed intriguing, but improbable.

Still, the professor was careful not to distance himself from McKane’s ongoing research and scientific publications.

McKane’s purpose in visiting the warehouse today is to address his mentor’s lingering doubts head-on. It is a rendezvous of two respectful and supportive colleagues. But in McKane’s gut this meeting feels more like a showdown. The inventor has implicit trust in his machine to replicate its mysterious process anywhere and at any time. Even at an abandoned boat house.

The question is whether he can win over the incredulous professor’s full confidence before he leaves.

* * *

Randy and McKane are in the boat-garage-turned-makeshift-laboratory for the day, both staring at the Quantum Kinetic Fusor, which stands unpacked and fully prepped on a timeworn workbench.

Although the professor has read the patents and the journal articles, he has never examined the device in a forensic setting. McKane aims to demonstrate the quantum machine in action while the professor watches.

Randy approaches the workbench to examine the setup.

"It's ready now," McKane says. "It's locked into resonance with the water. All you do to activate the process is turn the voltage knob." McKane looks at the professor. "Do you want to turn the knob?"

Randy hesitates. He did not expect direct involvement in the experiment. In fact, he expected the opposite since his duty as a scientist limits his active participation in a testing regime. Typically, he would establish experimental protocols for the test, then carefully observe the machine under study as a scientific witness.

But this case is different.

McKane is more than a client, more than a respected junior scientist. He is Randy's friend.

So, Randy accepts the challenge. He cautiously turns the voltage knob up.

From 0 volts to 10 volts.

From 10 volts to 20 volts.

From 20 volts to 30 volts.

From 30 volts to 40 volts.

Suddenly, Randy stops as if struck in the face by an invisible fist. He lets go of the attenuation knob and staggers back several feet. He leans against the opposite workbench.

"McKane, I have to tell you something," he says in a shaken voice. "I have been around a lot of powerful machines over the years. Some operate by huge electrical force. Some operate on nuclear power. I 'm not one to get physical reactions around machines. Matter of fact, I've only had two headaches in my entire life that I can remember. Right now, I have a headache."

Randy eases himself down to a sitting position on a battered old chair. He pauses a moment before asking a question.

"It's from the machine, isn't it?"

McKane reaches toward the machine and returns the voltage knob back to zero. Then he turns towards Randy and says, "Welcome to my universe. Now you know what it feels like to step inside a gravity well."

"I never expected this," the professor admits. "This power is amazing."

23

Race to Fusion

"It is going to be a race," Matt said.

"The first one to put a timestamp on the discovery of fusion wins in the minds of the public. That's all important. It's the whole ballgame."

Matt's prophetic words were spoken in their first zoom conference. His warning was unmistakable and ominous.

But there was more. Matt explained what "winning the race" means.

"You already have the data, everything we are looking for. The challenge is to convince the world what you have. Tell us what you have done. Show the world you have done something unequivocal in physics. Open Pandora's box. Put your science on the table. Publish your work in an important scientific journal. Confirm your discovery with simple scientific validation. Establish your economic propositions. Explain why you should be the one telling the story."

* * *

On December 13, 2022, McKane hears shocking news. In a surprise announcement during a press conference, US Secretary of Energy Jennifer Granholm proclaimed a major scientific breakthrough in fusion research at the Lawrence Livermore National Laboratory ("LLNL") in

California. According to Federal officials, LLNL researchers had successfully ignited a first-ever fusion reaction that propagated more energy than it took to begin the process.

News concerning the successful experiment is flooding the internet and dominating the leads on prime-time cable. After 70 years of research at Lawrence Livermore, the improbable has finally happened.

Official reports indicate the atom-smashing demonstration employed the largest array of lasers on earth, 192 in all, housed within a building three times the size of a football field, to blast a spec of deuterium until it exploded inside a containment vessel. Although the fusion event only lasted a billionth of a second, the accomplishment was being hailed as a major "breakthrough" and a "landmark achievement."

McKane thinks the breaking news impossible.

He sits down on a lab chair next to his library of old books. In need of distraction for a minute or two, McKane steals an upward glance as if to acknowledge these old companions—the books he has read and reread over a decade. Iconic books about gravity and electromagnetism. Biographies of James Clerk Maxwell, Oliver Heavyside, and Nikola Tesla.

Masterminds whose ideas changed the world.

Today, an astonished McKane wishes for inspiration from the animal spirits of these scientific mavericks of the past.

* * *

McKane is conflicted about the Lawrence Livermore news. He feels an initial rush of excitement over the apparent significance of the transformational experiment, portending good news for the planet and its inhabitants.

But he is also devastated. After all his effort and sacrifice, he will not be the first to place a timestamp on fusion.

He lost the race.

McKane scolds himself for not going public sooner when he first had his fusion invention in hand. He wonders why he had been so cautious by insisting on years of arduous research before breaking the good news in a worldwide press release. The delay cost him the opportunity to write his name in a mythical physics hall of fame. More importantly, the postponement cost precious years in the world's fight against the catastrophes of a warming planet.

The cascade of second-guesses forming in his mind deepens his unshakeable feeling of disappointment.

McKane closes his eyes, pausing the moment. He centers his thoughts on steady breathing, hoping to avoid descending the slippery slope of self-doubt.

When he opens his eyes, a sudden realization hits him.

No way! Too much diligence is the right amount of care before making such a tantalizing announcement to a waiting world. Wisdom dictates it better to be the stealthy tortoise rather than the sprinting hare, when chances are the public will freak out by the implications of such a far-reaching proclamation.

McKane is certain that his plodding methodology had been the right call all along. He had opened fusion's Pandora's box one inch at a time.

By first doing something unequivocal in physics.

By confirming the repeatability of transmutation at low voltage.

By establishing his intellectual property rights internationally.

By validating his discovery with scientific research at a national laboratory.

By publishing the dramatic experimental findings in a notable journal.

By revealing the vast commercial applications of fusion.

By extending fusion to the world now, rather than the distant future.

Despite these reassurances that he had charted the appropriate path toward ground-breaking innovation in fusion technology, McKane agonizes over the ultimate question: *Who should be telling the story of fusion*?

* * *

McKane needs a plan of action to get the ground-breaking news of his work outside of his inventor's bunker. To take his work public.

He must find the most adept person in the world who is currently writing about fusion technology. He must reach out to the best global data service company that works with IP owners. He must forge relationships with media empires to disseminate his story. He must contact the bipartisan U.S. Senate's fusion energy committee. He must dine with the governor.

It is time for a Press Release.

Stealth Laboratory Beats Livermore to Sustained Plasma Fusion Ignition

SEATTLE, Date, 2023, Quantum Kinetics Corporation, a company focused on becoming a leader in fusion technology, emerged from stealth mode in response to the U.S. Department of Energy (DOE) announcement of the achievement of fusion ignition at Lawrence Livermore National Laboratory ("LLNL").

Quantum Kinetics Corporation (QKC) is making public, for the first time, the existence of a modular, synthetic device capable of producing fusion plasma temperatures of over one hundred million degrees Kelvin (1 keV – 18.6 keV) at 1.9 – 4.4 watts input. The patented invention, as described by laboratory officials, has the potential to provide limitless, carbon-free electricity to the world – to be used to fuel and heal our planet.

Straight from the realms of science fiction, comic book tales and blockbuster movies, QKC has invented a real-life, functioning Arc Reactor - capable of producing controlled, scaled, and sustained nuclear fusion at room temperatures.

Inspired by a young American scientist who had been studying gravity and hydrogen densities in his garage for 11 years, QKC has successfully demonstrated safe and maintainable fusion – a technology that could permanently alter the course of human history. Executives of the company call the new process "Electro-Physical Transmutation" (E-PT™). The method works with distilled water, freshwater, seawater, and heavy water (D2O) dielectrics as its fuel source. E-PT™ operations can be fully vectored to meet any desired work requirements.

The company recently conducted research on its patented technology in a joint-strategic relationship with Pacific Northwest National Laboratories, a division of the U.S. Department of Energy. The research was published in <u>*NPJ Nature*</u> and heralded by the three journal reviewers as a "*much needed invention around the world.*"

For more than sixty years, nuclear fusion projects around the world have pursued the goal of producing abundant, clean energy by heating matter to more than a hundred million degrees until the superheated atoms melt into a magnetically confined plasma field, thus replicating conditions in the center of a star. This is the "star building" model to create nuclear fusion. Despite investments of billions of dollars into research and development, the goal of producing sustainable stellar-like fusion remains as-yet unsolved due to overwhelming complications and technical challenges.

Recently, December 5th, 2022, scientists at the Lawrence Livermore National Laboratory ("LLNL") in California hailed a successful fusion "ignition" experiment producing net energy gain. The demonstration employed 192 of the largest lasers on earth, housed within a building three times the size of a football field, to blast a spec of deuterium until it exploded inside a containment vessel. According to Federal scientists, the one-time event was a major "breakthrough" and a "landmark milestone achievement."

Despite the significance of the discovery, some scientists at Lawrence Livermore National Laboratory, including researcher and spokesman Alex Zylstra, cautioned that practical applications of fusion power remain a distant pursuit. Best estimates are that it may still take 20-50 years to realize fusion's commercial propositions as demonstrated at LLNL.

While LLNL was aggressively making strides of stellar-like fusion, a small U.S. energy company was cracking the low temperature fusion code -- bringing the idea out of the realm of myth and into the material world of capital markets and scientific measures – into a space where it can be used for the good of mankind and the earth we populate.

About Quantum Kinetics Corporation (QKC)

QKC holds patents both nationally and internationally on fusion devices. Company officials assess its machine's capability to produce energy gains over losses is attributed to low voltage synthesis of heavy atomic nano-elements. The ET process is also distinguished by Biological Index Stimulation of unique diatoms and life structures previously unknown to science.

Recent economic studies estimate nuclear fusion to be a multi-trillion-dollar industry. As such, it is projected to be one of the largest, if not *the* largest, industry in the world.

QKC projects its revolutionary work with fusion energy will provide for the world's poor and under-resourced. And it will open countless doors into a future touched by new hope and healing – and someday a gateway to the stars.

For Further Information visit: http://www.quantumkinetics.co

***

McKane is convinced that from now on he must take control of the story. There's no more time to spare. Getting an article published in a prestigious journal, coupled with a national press release, is a great start. But commercialization of an unprecedented scientific discovery is only bolstered by massive publicity.

His invention needs a wider audience. And a brand.

McKane decides to up the ante by creating a "Quantum Kinetic Company" website and an accompanying YouTube video to take his breakthrough technology worldwide. The time has come to enter the larger conversation and put his invention before the eyes of the public.

He understands the consequences of stepping onto the public stage all too well. Loss of anonymity. A storm of controversy in the academic community. Public quarrels about scientific validity.

He expects a barrage of thinly veiled accusations on social media that his emergent technology is just another phony 'get rich' scheme. And he's already bracing for the widespread insinuations from the energy sectors that his story is 'too good to be true.' Stalkers and detractors will come out of the woodwork the moment he pushes his work into the public eye. It's to be expected, and these are serious costs to consider.

However, McKane addresses his own internal Socratic debate about this public strategy with a non sequitur. *The essence of 'cold fusion' is technology so amazing it initially startles the intellect. To doubt such news at first is a natural response. But people will eventually warm up to the idea. They simply need time.*

And McKane has none left to spare.

Convinced of this fact, he jumps headlong into the fray.

24

Friends Into Family

McKane's background as an athlete helped form his vision of the role of teammates in achieving personal success. Stirred on by teammates, the gymnast who falls off the rings gets up and tries again until the skill becomes second nature. And when the pole-vaulter takes the inevitable hard plummet from a height above 18 feet, teammates press even harder.

Victory, in short, is never achieved alone.

Now, on the precipice of unveiling his breakthrough inventions to the world, McKane leans on the constant encouragement from his team of supporters. As an inventor, he's re-learning the athletes' creed: *The flipside of every failure is an inspiration to keep trying.*

* * *

McKane remembers the drive to Kerri's house.

It is a June afternoon, and it is sunny. The route takes him through an affluent neighborhood with upscale homes on expansive, manicured acreages. He is enjoying the scenery and leisurely drive in his classic 32-year-old, hydrogen-hybrid BMW. The car still gets its share of thumbs up from other drivers. He welcomes the attention as if each wave signifies the universal acknowledgement that *they don't make them the way they used to.*

He doesn't know how many guests will be attending Kerri's dinner party.

Some friends will be there, along with a few executive retirees from Boeing. He has never met these former administrators of the aeronautical giant. Butler, his honorary grandfather, will be there too. In fact, the party was Butler's idea, another opportunity for his adopted grandson to spread the word about a new form of plasma energy.

It is going to be the first time McKane meets with a group of potential investors. He wonders if the meeting will bring any serious attention to his technology company. He thinks about making a formal presentation to the group but quickly dismisses the notion.

All he wants to do is enjoy an evening with supportive friends and savor a wonderful meal prepared by Kerri herself. It should be something special – after all she is the owner of several fine dining establishments in the Pacific Northwest.

He arrives 30 minutes late but is certain Kerri won't mind, believing she intends their meeting to be a casual and open-ended affair. He pulls his "Bimmer" into a parking spot next a new M car. He wonders who owns the European Autobahn-inspired race car.

McKane is not used to arriving late. He makes a courteous apology to Kerri at the front door, who waves it off as unnecessary. Without hesitation or comment, she ushers him into an open and appealing patio area. Other guests are already assembled around a large dining table. Surrounding the main table are smaller tables offering an abundance of hors d'oeuvres and open bottles of imported Italian wine. The engraved wine glasses are neatly positioned at each place setting.

He sees there are only two remaining empty seats. Both are at the center of the dining table. Adjacent seats.

He takes one of the remaining chairs. Moments later, the final guest arrives as if timed perfectly by the hostess. He takes the seat next to McKane.

Butler makes the introduction. His name is Bob. He spent a long and successful career as an aeronautical engineer at the Boeing Company. He has seen it all. Done it all. From start to finish, he knows how to get big things off the ground and flying.

Butler coyly asks Bob if he read the journal article sent to him earlier in the week, the paper about "Electro-Physical Transmutation"?

Bob smiles. "Of course, I did. It is fascinating. There are so many questions to ask the author. Whoever wrote it must be quite intelligent."

Butler pauses, sensing the cast of characters is now properly assembled and the audience quietly expectant. With perfect timing, he makes the dramatic revelation as if announcing, *curtains up!*

"Lucky for you the author is here with us today. He is the gentleman sitting next to you!"

The surprise announcement draws an exclamation from Bob. "OMG. He is so young! I would never have suspected that!"

* * *

Bob asks his questions for the next two hours. From center stage, McKane does his best to answer. The discussion is far-ranging. It captivates the larger audience listening in. Soon, Bob poses a direct and unexpected question, "How much money do you need?"

* * *

One of the guests offers to take McKane for a ride in his M car before he leaves the party. Overhearing the offer, Bob follows them to the parking area. But there is only room for two in the race car. Bob grabs the driver's arm and whispers in his ear.

"Be careful. Your passenger is the most important person in the world."

25

What Many Hands Make

McKane did the impossible – stimulate something living out of something lifeless. By making the first real-life and fully functional "Genesis Device," he proved Lord Byron's 200-year-old proverb to be true.

Life is stranger than fiction.

And there's more. His amazing machine crosses the impassable barrier between inorganic and organic chemistry, while at the same time releasing unlimited power from nature's own free-energy domain. Despite accomplishing the improbable, McKane worries,

People will say it is too good to be true.

Yet, the research shows each D2O test produces organic carbon from 99.9% pure, inorganic deuterated water. The CDOM non-filterable organic carbon increased 1,500%, according to the National laboratory. Worm-like structures appeared spontaneously and then transitioned through complete life cycles within 24 hours. The non-filterable organic material appeared in each test, with the 30-volt treatment stimulating the largest population of organic carbon.

With each tranche of confirming evidence, the entire world is becoming richer, revitalized. But only if

Seeing is believing.

McKane postulates the staged 30-60-90-volt treatments captured element fusion at various transitional stages of transmutation. The evidence shows the volume of fused substances is related to the intensity of the applied voltage.

He sees the E-PT™ process akin to organic alchemy coupled with inorganic metallurgy. No wonder the national science team was taken aback at their initial analysis of post-treatment D2O samples. They ran re-tests to confirm the findings. Now the team is standing strong behind the data and publishing the results on the DOE and PNNL websites.

McKane did not accomplish the impossible alone.

***

McKane considers the long cast of supporters, antagonists, skeptics, partners, and friends who played a role in his improbable journey as a lost artist, a broken firefighter, a desperate dreamer, and a revolutionary inventor.

He thanks God most of all for his father, Riley—McKane's greatest supporter.

And for Butler, Matt, Nick, Roger Emerson, Dr. Pollack, Randy, and Scott Braswell among others. For the kind widow who rented him the garage where his invention was born.

So many voices injecting hope into a bleak world. Animating his work. Lighting his path. Propelling him to success.

And obstacles, too. Cynics lined up against him. Odds stacked the wrong way at so many junctures. Heartbreaks built on the ruins of other heartbreaks.

And yet they form and shape his story, like pressure and heat bringing diamonds out of coal. Somehow, he's grateful for these, too.

McKane finds himself on the brink of a fulfillment of a dream. A product of heartbreak and healing. Aiming where he has no business to aim and dreaming something he has no business to dream.

Somehow, he's cracked the code. Brought cold fusion out of the realm of myth and into the material world of capital markets and scientific measures—into a space where it can be used to heal the planet.

His work will provide for the poor and under-resourced. It will open countless doors into a future touched by new hope and healing.

It took an outsider to crack that code. An unconventional coach and athlete. A student dismissed by engineering classmates and industry scientists for what they deemed a lack of respectable credentials.

Nevertheless, he cracked the code humanity was unwittingly waiting for someone to step forward and solve.

He pulled the sword out of the enchanted stone.

And all it took was an artist.

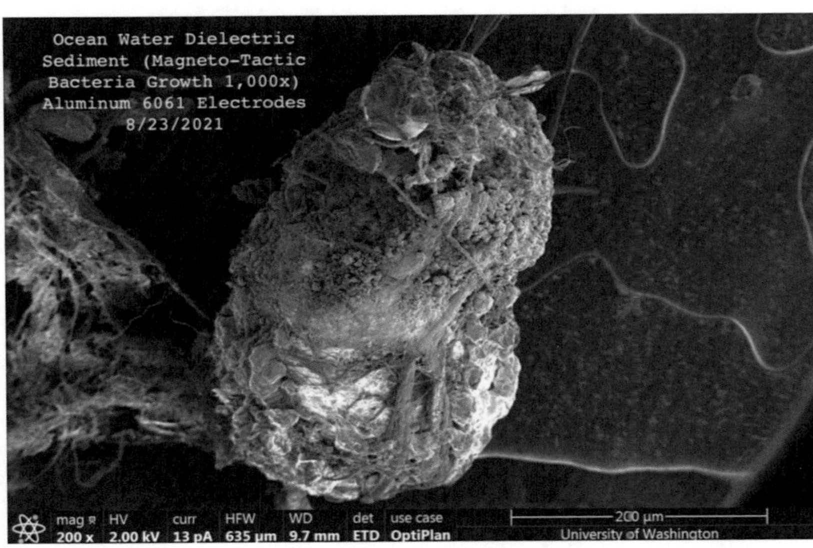

Magneto-tactic Bacteria 1,000x growth post treatment – 8hrs

26

At the Edge of Paradise

It is the third Friday of February. Outside the lab, temperatures have fallen below 20 degrees for the third day in a row. A blanket of ice and snow is thickening by the minute. Soon, McKane will be landlocked on a steep hill with outside temperatures that could kill an unprepared man in minutes.

With a glance at the treacherous snowscape outside, his mind slips back to the day of the accident.

This Subaru revolving in mid-air. The desperate screams. The crawl of each moment as he hung helpless between life and death . . .

McKane has not allowed himself to remember Happy for years. At least not as anything more than a concept. It seems a lifetime ago, before the fateful road trip along a treacherous mountain corridor. But this frigid and lonely morning is different.

They had been married for two years and still blissfully lost in the wonder of an improbable romance. Their home was on Sioux Drive—a fitting address, given McKane's ancestral Indian bloodline. One day he hoped to pass that ancestry on his mother's side to children.

To his and Happy's children.

He reflects on their initial plans for a family together, the prime purpose for their union. He recalls an old and cold decision. One that still haunts him from time to time.

Perhaps it wasn't too soon for children after all.

He can still hear the echoes of Happy's tired admonitions, scolding him for spending too much time on his experimental, hydrogen-powered car out in the garage. But she espoused a probability he could never accept.

It's not going to work.

***

An incoming email grabs McKane's attention—an important message from Ohio is coming through. It is Roger Emerson, confirming the acceptance by the USPTO of his "Quantum Injector" patent—the water-based fuel technology to revolutionize and retrofit internal combustion engines everywhere.

With this patent, the game has officially changed for automobile technology. For his invention generates explosive energy instantly from simple water mist. It is the mythical "repulsor."

Water Mist Injector (Repulsor Technology)

McKane is more than astonished by the swiftness of the award, only eight months. He had secretly wondered if the Patent Office would even endorse his novel energy technology—a source of power surpassing both gasoline and diesel.

More readily available in Nature's storehouse.

More economical to serve the disenfranchised of the world.

Potent as a tool to destroy the specter of carbon build-up in the world's atmosphere.

He will call it "Star in a Car." But the applications are limitless. The tech can be retrofitted into any internal combustion engine, including massive machines and even jet planes.

McKane can't come out of the garage now—not even for a minute.

***

McKane is on a two-week vacation in Kauai. His father, the psychologist, had finally persuaded him to step out of his inventor's bunker. It would be a "mental health" holiday, explained his father—time enough to recharge his creative energy reserves with long hikes into remote landscapes, endless beaches, and waterfalls.

The barefoot existence would be the perfect counter measure to the loneliness of the lab, the pressure of verifying a new technology, and the burden of sharing it with the world.

***

He is on a long hike that lasts the day. On the return leg, he passes a young lady sitting alone on a ledge overlooking the ocean. She appears to be writing something in a small book. They acknowledge each other with brief "hellos." As he takes the next turn in the trail, he stops in his tracks. He suddenly remembers the terrible things that can happen on the edge of a cliff.

Life-ending things.

He quickly starts back, hoping to find the young lady still perched at her lookout. When he arrives, she is still writing.

It is an awkward moment. He realizes she may feel threatened and race away. He quickly tries to explain himself.

"Hello again. Sorry, but I just walked past the most beautiful woman in the world. I had to return to tell you."

The young woman smiles, amused by the transparent flattery from a handsome stranger. It is clear she will not be fleeing on the reverse footpath. Instead, she introduces herself as "Sommer." She explains she is a doctor on the island. She often comes to the ocean observation post to write in her diary.

She is curious about him. "Are you a visitor or an islander?"

"I'm sorry to say just a tourist," he replies. "Maybe someday that will change."

"What do you do for work?" she asks.

"I do research on water and energy."

Pointing to the ocean below and to an endless blue horizon, he says, "I can turn a cup of seawater into a synthetic meteoroid, then use the process to generate endless clean energy."

Amazed or in disbelief, she asks, "How can you do that?"

"It's quite a story," he replies.

END

The Portal at the Edge of Paradise

The Quantum Kinetic Chronicles

Episode II

Journey Beyond the Quantum Horizon

1

Atomic Field Trip

It was like when I was on the soccer field and wanted to play a position that the other kids didn't. I learned that there was more of a risk but also more of a reward if you ventured into a new area.

Jennifer Doudna,
The Code Breaker — Gene Editing and the Future of the Human Race

Don't worry dad, it's going to work.

McKane quoted himself for the thousandth time, repeating a phrase he often used when attempting the difficult. These words had become a soundtrack for the boldest steps toward accomplishment in his life. He returned to them when setting another record in the pole vault or gaining admittance to a prestigious graduate program in a foreign country.

This time, he spoke them as he attempted something unprecedented in physics.

McKane and his dad were in Randy's engineering lab late in the evening. Earlier that night, they had picked up dinner at a local sandwich shop, knowing their work with the Scanning Electron Microscope (SEM) might go past midnight. Their objective was to analyze the atomic specimens from recent tests with the Quantum Kinetic Well (QKW).

Randy — a university professor, national expert in electrolysis, and relentless forensic investigator — found the images unnerving.

Analysis Report: Image32-4

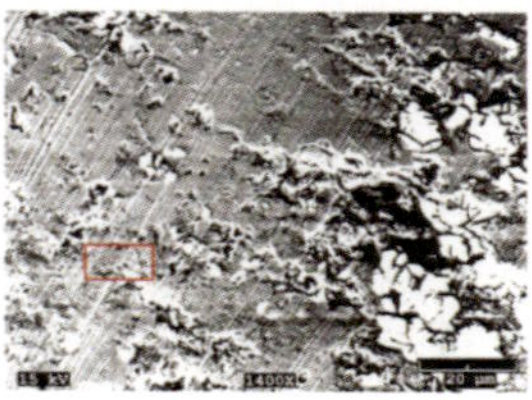

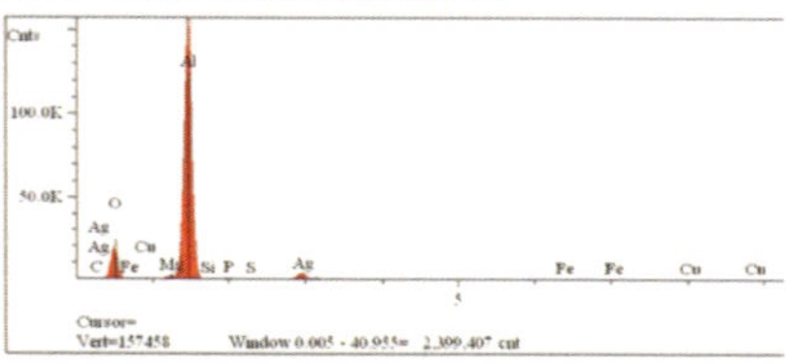

Elt.	Line	Intensity (c/s)	Conc	Units	Error 2-sig	MDL 3-sig	
C	Ka	14.78	0.343	wt.%	0.153	0.228	
O	Ka	4,453.52	22.315	wt.%	0.127	0.052	
Mg	Ka	713.77	0.942	wt.%	0.019	0.022	
Al	Ka	49,984.19	68.126	wt.%	0.112	0.022	
Si	Ka	124.92	0.273	wt.%	0.024	0.034	
P	Ka	270.24	0.560	wt.%	0.023	0.029	
S	Ka	75.49	0.153	wt.%	0.019	0.028	
Fe	Ka	69.31	0.472	wt.%	0.047	0.065	
Cu	Ka	34.56	0.514	wt.%	0.081	0.113	
Ag	La	778.08	6.302	wt.%	0.113	0.118	
			100.000	wt.%			Total

kV 15.0
Takeoff Angle 45.0°
Elapsed Livetime 30.0

Analysis Report: Image30-1

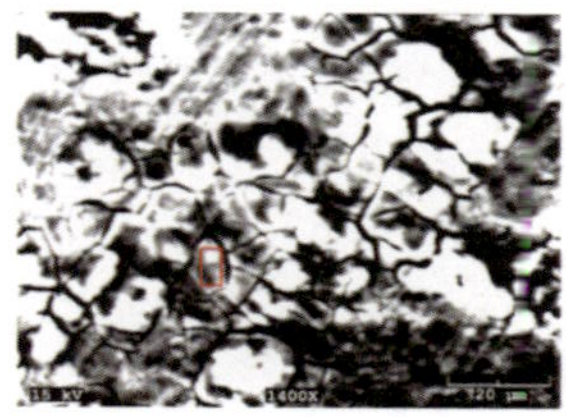

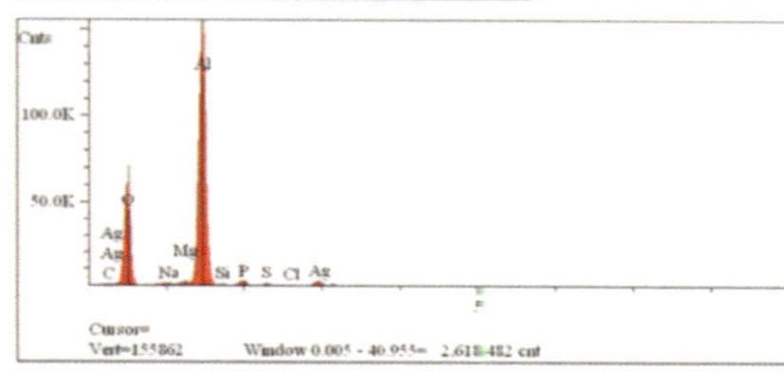

Elt.	Line	Intensity (c/s)	Conc	Units	Error 2-sig	MDL 3-sig	
C	Ka	104.49	1.425	wt.%	0.108	0.145	
O	Ka	13,012.90	42.291	wt.%	0.137	0.032	
Na	Ka	413.31	0.547	wt.%	0.015	0.017	
Mg	Ka	738.34	0.767	wt.%	0.014	0.014	
Al	Ka	47,806.49	48.879	wt.%	0.082	0.015	
Si	Ka	301.72	0.431	wt.%	0.017	0.022	
P	Ka	876.37	1.217	wt.%	0.021	0.022	
S	Ka	507.37	0.706	wt.%	0.018	0.021	
Cl	Ka	94.77	0.137	wt.%	0.015	0.022	
Ag	La	632.52	3.600	wt.%	0.077	0.086	
			100.000	wt.%			Total

kV 15.0
Takeoff Angle 45.0°
Elapsed Livetime 30.0

FIGURE 1: EDS & SEM of aluminum (Al) electrodes post treatment of well water showing the emission spectrum spike of Silver (Ag) growth from Electro-Physical Transmutation (E-PT). The Quantum Kinetic Well (QKW) phenomenon.

"This is amazing. Your machine reduces Strontium with very small injections of power. And look! There is silver on the electrode's surface. I've never seen or heard of anything like this. We might have found a viable commercial application of your invention."

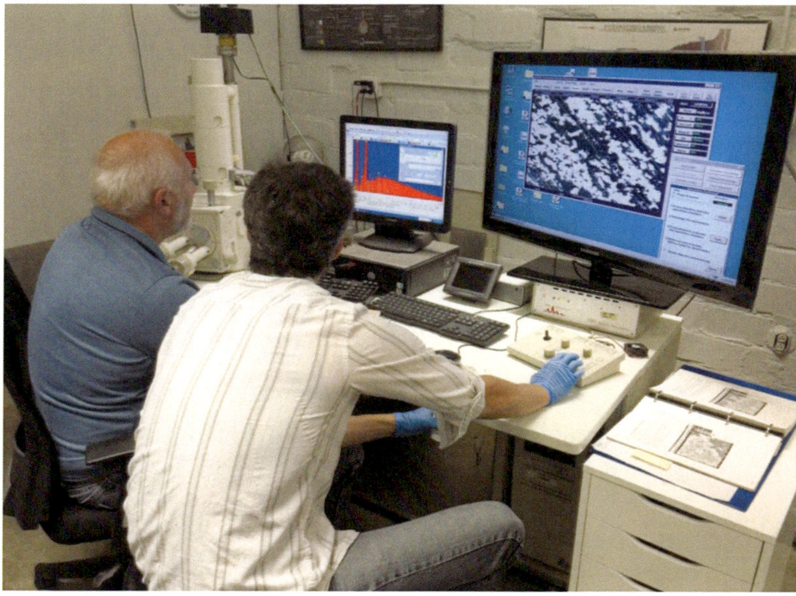

FIGURE 2: With the aid of a Scanning Electron Microscope (SEM), Randy (left) and McKane (right) analyze the post-treatment sediment fused onto the aluminum (Al) electrodes.

* * *

After the SEM revelation, the Quantum Kinetic team believed they had discovered a possible solution to the non-permanent storage of used nuclear fuel at Hanford in Washington state. They reached out to officials at the Hanford nuclear site and offered their help, assuming they would be welcomed. Instead, they encountered a bureaucratic blockade as mighty as a fortified medieval castle — the kind with moats and armed garrisons to keep out intruders.

Following two months of ghosted emails and ignored voicemails to the team at Hanford, the men from Quantum Kinetics devised a new plan to circumvent the barricade. Rather than beating on the front door, the Quantum Kinetics team would use Randy's clout as a national expert to get them in touch with a department head. Randy might help them open a back door.

The plan worked.

In a telephonic conference with a Hanford production manager and his assistant, the researchers from Quantum Kinetics described their successful water-cleansing invention. McKane showed them how his invention had easily destroyed Strontium, Cesium and other contaminating nuclides known to cause cancer while at the same time holding a record-setting 10 kiloelectronvolts X-ray photons or 100-million degrees Celsius for 8-hrs at only 1.9 watts. The revolutionary achievement had already been published in Nature's partner journal, *NPJ Clean Water*.

Despite this remarkable disclosure of what McKane's vaunted invention could do, of its obvious economic and ecologic promise, the Hanford program manager was not interested.

McKane remembered his words well.

"Good morning, my name is Sexton. Glad to meet everyone. I am here with Vivian, a chemist at Hanford. I think Randy has done some work for us in the past, and we know a lot of the same people. That is why we are willing to talk today. By way of an update, I am not in clean up at Hanford. I am the deputy project manager for tritium production. I oversee science and technology efforts to support the production side of tritium for the strategic stockpile and for national defense. I am not connected to controlling tritium leaks."

McKane wondered if they had been connected with the right department.

Then Vivian introduced herself. "Hello, I am an inorganic chemist at Hanford. I study the behavior of tritium as a hydro-species and its sub-micro measurement in tritiated wastewater and the hydrosphere. Sexton is more on the nuclear reactor side."

Sexton agreed. "That's right. I work with design teams focusing on T rods and the best mechanisms for harvesting tritium. The T rods were built by Westinghouse in South Carolina. They have been in production for more than 20 years. The rods produce one gram of tritium per rod over 18 months, when lithium absorbs a neutron. The tritium is stored in solid state, not in the rods. In the core of the nuclear reactor, tritium's issue is soft beta, but there are also a lot of gammas that are hot, extremely hot."

McKane appreciated the education about tritium production, but thought Sexton was avoiding the real reason they had secured this conference — to discuss how to reduce used nuclear waste.

Sexton continued his lengthy soliloquy, weaving further off topic to describe frequent travels to Tennessee and its four nuclear plants operated by the Tennessee Valley Authority ("TVA"). He explained that the TVA provided approximately half of that state's electrical power.

"I have important work with the TVA. They know me as the national expert in tritium production," Sexton bragged.

Not wanting to offend his host, McKane let his mind wander as Sexton prattled on. *I agree about nuclear fission as a great way to stream unlimited carbon free energy onto the grid, but what about safeguards for the people within a 50-mile radius of the nuclear plants? What about safe farmland, safe drinking water, safe milk for the infants…why can't we have both unlimited clean energy and safe nuclear?"*

Sexton finally reached his point. "I have plans to retire soon. I am moving to Texas. I don't want to spoil my legacy by getting involved with your project. Anyway, you can't turn a proton into a neutron synthetically. And a low-power device like yours can't alter any isotopes. That's impossible."

Frustrated by Sexton's dismissive comment, McKane nevertheless held his return fire, not seeking an argument — only collaboration. It was clear Sexton had never read the QKC's published research and expository white papers.

Vivian interrupted, checking her superior's dismissive comments and re-focusing the conversation. "On the other hand, I understand what you gentlemen really want to know is whether we can come up with a means to collaborate on solutions. Your technology sounds good, but I would like to see your device scaled up substantially. If we can do that, then this will be a great application here to clean up the ground water problems. Overall, I see your technology will also reduce the strain on the reactor by making things cleaner. We have the capacity here to do more studies with you. To begin, we might consider integrating your device as a secondary to our primary clean-up process. Maybe this is the best path forward to begin our work. I also have someone in mind who will be very interested in your technology. This may be a revolutionary solution to a big problem."

The Hanford scientists wrapped the phone conference with the promise to keep in touch. And in good faith the Quantum Kinetics team attempted to follow up with Sexton and Vivian many times after the call.

But as before, they encountered an impenetrable blockade.

This time, both the front and back doors of Hanford were securely locked. But the team was not content to treat their dismissal as a dead end. There were other ways in.

* * *

The drive to Hanford was long and tedious. Garry, the manager of an Everett machine shop, was driving the car. McKane was his front-seat passenger.

Butler, now only weighing 122-pounds, was positioned in the back seat by himself. He needed the space to stretch his legs diagonally to assuage his chronic back problems. Butler was also having trouble breathing, a chronic symptom of severe CPD from decades of unprotected work in his machine shop. The doctors said he was running out of time.

Butler was grumbling.

Is this test on the river water going to work?

Are we wasting our time?

Will it be raining all the way to the Tri-Cities?

When can we take a break?

FIGURE 3: McKane (left), Butler (middle), and Garry (right) driving to the extraction point to grab some river water near Hanford for transmutation testing.

Garry and McKane were sympathetic. They reassured their good friend with patient answers to each of his questions. It was obvious that he was in constant pain and struggling as a cooped-up passenger on a long road trip. But it was in fact Butler's choice to accompany his buddies on this drive, despite the steep costs to his frail and failing body. After all, Butler's final purpose in life was to support his adopted grandson, McKane, on his breakwater scientific quest.

The sky finally cleared as they reached the summit of the mountain pass. It was a sign that they had transitioned out of the heavy rain showers on the west side. Soon, the sun would come out and the snowy landscape would be blinding. They were making good headway towards eastern Washington and their daunting destination — Hanford, the most radioactively contaminated place on earth.

Beyond the mountain pass, the travelers found themselves in an endless flatland, a virtual desert. Butler was bored. He engaged Garry and McKane in conversation to distract him from the pain. He peppered his friends with questions about what they would do when they arrived at the river near the nuclear plant.

How will we find the best location?

How close can we get to the boundary?

How will you collect the water samples?

Is it safe to get into the river?

Will we be surveilled by federal agents?

What's going to happen if we accidentally enter the restricted area?

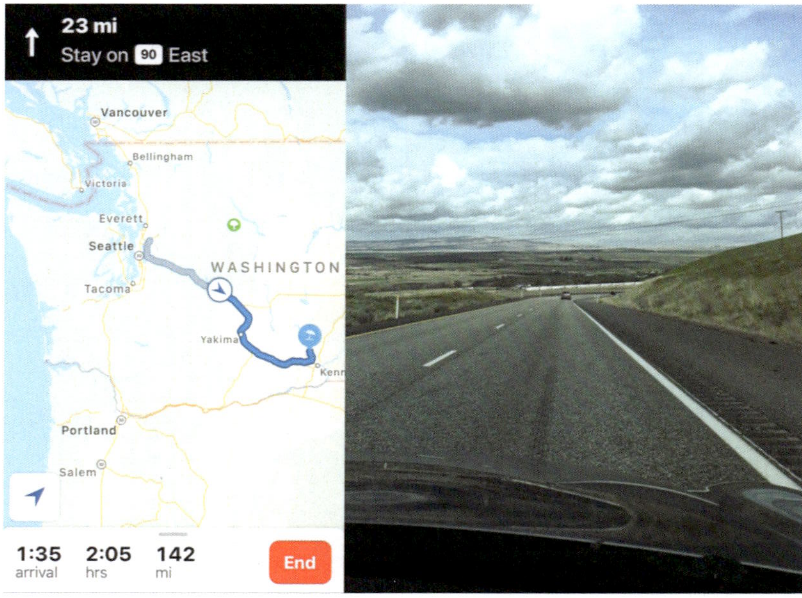

FIGURE 4: McKane's iPhone Google Maps plotted the course to the Hanford extraction point.

McKane answered Butler's nervous inquiries in clipped, but reassuring, sentences.

Don't worry Butler, it's going to work.

Getting the water samples will be easy.

Faster than going through the drive through at Starbucks.

The tests that follow will be harder, much harder.

Hang in there, pal.

***

McKane's iPhone indicated they had almost reached their destination. Garry stopped the car on the road leading to Carbody Beach, a popular recreational area during Covid.

It was now little more than a public dumping ground.

They saw a large sign announcing "No Entry at This Time" to the area, part of the local authorities' attempts to clean up the polluted beach. A heavy chain blocked public access to vehicles.

Frustrated, the friends quickly devised an alternate plan of action. They would backtrack to a point just south of Johnson Island. There was a public shooting area nearby called the "Extraction." It was directly downstream from the nuclear storage area and would suffice for their work.

FIGURE 5: The extraction area near the Hanford nuclear site.

The moment they arrived, McKane got out of the car and stood motionless, surveying the foreboding shoreline. At his feet, the dark soil was smattered with the carcasses of birds, insects, and freshwater crustaceans. He noticed the offensive odor of rotting flesh. There were threads of brown, lifeless river grass wafting in the current near shore.

McKane's mind went numb as he absorbed the bleak tableau confronting him downstream of the nuclear plant, like a fragile nerve exposed to open air. His breath tightened as a stark realization sunk in. *The local lore about dogs dying after drinking the water here may be true after all.*

Reminded of the seriousness of his purpose, McKane turned his focus to getting the water samples and getting out. He studied the river to estimate its depth. He asked his companions for help to unpack the tools and carboy. Butler and Garry stayed close to the car, not wanting to venture towards the water. On the higher ground, they could also keep a better lookout for strangers.

As McKane moved the equipment down to the waterline, a small boat drifting westerly towards the city near the opposite riverbank suddenly changed course and veered towards the researchers' position. The lone occupant was not fishing for the last bite of a migrating salmon.

He was surveilling the three gentlemen on the shoreline.

Butler called, "Hey look out. That guy in the boat is checking us out!"

"This is weird, McKane," Garry yelled. "Better hurry."

McKane shed his shoes and put on heavy boots, latex gloves, and protective eyewear. Without hesitation, he waded into the river with carboy in hand while Garry and Butler readied for a confrontation—or else a quick getaway.

***

McKane planned to treat the Columbia River water three times with his electron bifurcating machine. Each trial would be run for 12 hours at separately staged voltages — 30, 60, and 90 volts, respectively. The total power consumed in each test would be less than the amount of energy needed to recharge a cell phone. His experiments would take place at the oceanside laboratory, a location he preferred because of its remoteness. McKane invited his father, the retired psychologist.

Do you want to have three days down at the beach lab?

I could give you a refresher course on running the equipment.

You need to know in case something happens to me.

Maybe we could find time to play golf in-between run times.

FIGURE 6: The carboy containing the extracted water ready for transport to the remote lab near the Pacific Ocean.

2

Archimedes Out of Time

Give me a lever and a place to stand, and I will move the earth.

Archimedes

It was the second day of May 2023. McKane and his father were walking easterly on a paved trail stretching along the riverfront and into the historic city. Visible ahead lay Tongue Point, the campsite of the Lewis and Clark Expedition during March 1806.

Despite the significance of the location and the magnificence of the Columbia River estuary, their immediate concern was not sightseeing. For three days they had been conducting scientific experiments in a tireless search for an ancient secret bound up in electronics and physics.

Their aim, which so many along the way had deemed impossible, was to stop climate catastrophe by uncovering an unexpected truth about the universe.

They had just finished lunch at a popular restaurant, the *Bridgewater*. The famed bistro had been unusually crowded. Disembarked passengers from a cruise ship docked nearby had packed out the restaurant like sardines. Their ship, the *Ruby Princess,* with accommodations to hold

3,080, was destined for Alaska. But her first stop was the oldest settlement in the United States west of the Rocky Mountains — Astoria — founded by Jacob Astor in 1811.

The researchers' route along the northwest section of the riverside foot path led first to Maritime Memorial Park. McKane's truck was not far away, parked under a soaring superstructure, the four-mile bridge connecting the states of Washington and Oregon.

Curiously, for the past three days a dark sedan emblemed *"Property of the U.S. Government"* had also been positioned in the same parking area within feet of the researcher's Ford pickup. The reason for the continued presence of the mysterious black sedan was unknown.

Beneath the looming shadows of the Astoria-Megler Bridge, a sweeping quarter circle wall of architectural concrete extended from the river's edge towards the center of the peaceful park. It looked like a sturdy branch on a sequoia tree. The expansive display was the "Memorial" itself, listing hundreds of names of those who lived and died near the Columbia River at its convergence with the mighty Pacific Ocean. The point where they merged formed the most treacherous harbor entrance in the world.

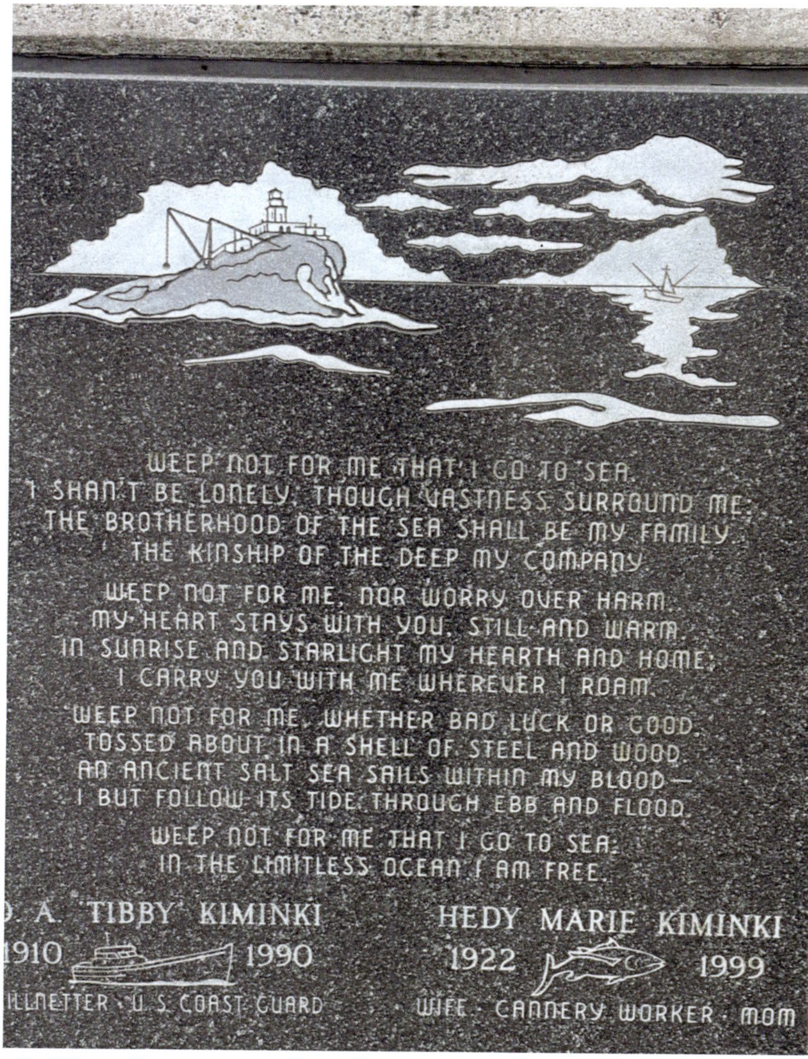

FIGURE 1: A commemorative tablet at the entrance to the Maritime Memorial Park in Astoria, Oregon.

As the two proceeded into Memorial square, a lone figure riding a rickety red bike entered from the opposite side. The bike rider was heading on a beeline for father and son, as if he had an urgent message to deliver.

They met the biker in the center of the courtyard.

FIGURE 2: The courtyard of Astoria's Maritime Memorial Park (shown as of May 7, 2025). This is the location where McKane and Riley interacted with the stranger on May 2, 2023. The area was an open-air green space in 2023. It has now been upgraded by addition of a reclaimed historic circular monument positioned inside a ring of discrete circles.

* * *

Legends hold that Archimedes was an ancient wizard with the superhuman power to lift ships out the harbor with a wave of his hand. But the true saga of Archimedes, who lived in the third Century BC, was stranger than any Greek fable ever told. Well-documented accounts by contemporaries revealed Archimedes did not need magic or alien powers to change European history forever. And when Romans invaded his beloved city of Syracuse, Archimedes employed his own ingenious inventions to do things undreamed of:

To raise foreign warships into the air.

To smash military watercraft to pieces with a giant hand.

To launch deadly projectiles at enemy ships with an ingenious catapult system.

To deploy a *Death Ray* to incinerate the invader's battle craft.

Archimedes' talent for invention did not stop with amazing military tools. He was also known for his many practical scientific discoveries and mathematical insights. His most famous invention was *Archimedes' Screw*, a clever device for systematically raising water levels within a contained chamber, a methodology still in current use for agricultural irrigation and wastewater treatment systems worldwide.

Archimedes also first realized and refined operational principles of physics now used in everyday work by engineers and scientists. His wonderful theories included the buoyancy of water, the center of gravity, the calculation of the area of a circle utilizing the approximate value of *pi (Π)*, and the law of levers and pulleys—the theory and practice of moving heavy objects with small forces.

Archimedes' fame as inventor and mathematician spread from Greece across the civilized world during his lifetime. He was sought after as consultant to kings and empires, solving complex problems with simple machines and practical solutions.

Greek history records that Archimedes lived three quarters of a century, relentlessly pushing the boundaries of science onto untrodden ground — until his final day. His famous last words echo down through history, revealing the inventor's uppermost attachment to his work product.

He had been at his home working with a new visual aid to solve another complex math problem when an invading Roman soldier stealthily entered his house. As the cowardly villain raised his sword to assassinate the genius, Archimedes' first thought was to protect his intellectual property:

Do not disturb my circles, he demanded, referring to his technical sketches and illustrations, as the assassin's steel blade plunged into his flesh.

* * *

In another marine harbor in another era of discovery, a cyclist's trajectory placed him directly in front of two pedestrians.

The stranger was moving fast and gave no indication he intended to veer from his course. McKane and his father braced themselves, the way a driver might do on a two-way street when confronted with an imminent head-on collision.

At the last second, the intruder braked and swiveled his ride sideways, blocking the footpath. He stood tall astride his bicycle, gazing intently at the gentlemen before him. His gray hair was disheveled, freely flowing in the sea breeze. He had on wrinkled denim jeans and a grungy wool shirt, proof his scruffy clothes had not been washed for weeks. The itinerant's feet were shod with tattered leather sandals. Clearly, the curious visitor had other things on his mind rather than care for the details of his appearance.

"Excuse me, gentlemen. Sorry to bother you. But I'm here to help you," he said.

McKane's father glanced at his son and then at his wristwatch, calculating the odds the chance encounter would turn out to be a waste of their time. But as a psychologist and a consummate professional, he wanted to hear more from the man before making his judgment. For the moment, he would act the familiar part of a seasoned clinician studying an odd patient for the first time.

The biker continued. "Do you see all these people crowding the path into town? I won't talk to them. Most of them are paranoid and suspicious of foreigners. They put their heads down and walk away from me, even before I say a word. You two are different. I know you are not paranoid. You will hear me out."

The psychologist's interest was piqued. After all, he had been clinically analyzing paranoid people for 50 years.

"Don't worry," blurted the old-timer on the bike. "I am not here to sell you anything or upset you. I know who you are. I knew before you arrived."

McKane's father nervously side-glanced at the hundreds of tourists moving in unison into the city, like columns of soldiers on an important mission. He imagined their real purpose was less ominous than this strange man assumed. They were buying souvenirs in novelty shops or perhaps catching a tasty espresso at the Fourth Street Cafe.

He turned his attention back to the fascinating stranger. "Okay, we are listening."

"I need to tell you about something important that happened here back in the 1800s. First, I have a question. Do you know what *Glendonite* is?"

"Not specifically," McKane replied. "But I can guess. Does Glendonite come from the ocean?"

"Yes."

"Is it calcium-based?"

"Yes."

"And does it grow in the ocean's colder waters?"

"Matter of fact, it does. How did you know that?"

"It was a lucky guess," McKane said with a wink.

The quirky man smiled back at him. "I knew it. You already know!"

Satisfied with McKane's answer, the biker's mood suddenly intensified. "I am here to ask you not to overlook this region of the

Columbia River in what you are doing right now. Your work is more important than you realize."

Stunned by the foreigner's awareness of their discreet investigation into the Columbia's water near Hanford, father and son remained silent. They stared back at the man.

"They say a large ship went down at the harbor entrance more than a hundred years ago. You can see a model of it down at the Maritime Museum. The vessel had been collecting important artifacts from other civilizations around the world. After sinking, much of the craft's contents were lost, but some remnants washed ashore or were carried upstream on the estuaries' strong tides. A few of the artifacts were eventually deposited upriver. I suggest you don't fail to investigate this material going forward. It will be important in your work."

McKane's father said, "I am wondering if you do work in geology or oceanography at the university level."

"No, never. I am retired."

"Retired from what kind of work?

"I was a sawyer."

"Okay. Can you show me your hands?" asked the psychologist.

The man extended his hands for inspection, as if complying with a detective's order under interrogation.

"I don't believe you about being a sawyer," the psychologist said. "You have no missing fingers."

"Right. I can tell you are a very perceptive person," replied the old cyclist. "To better answer your question, I was a carpenter. I lived with my family for 30 years. We had a business making furniture out of native wood."

The stranger abruptly changed the subject, as if he had unintentionally disclosed a family secret.

Veering the conversation back to his original subject, he said, "There are river caves and an abandoned city nearby. I have the keys. You can have my help, but only if you ask for it."

He spoke in cryptic phrases, like a misunderstood prophet.

The clinical psychologist fell silent, at a loss for what to say to this stranger.

McKane was nonplussed. Nevertheless, he showed some interest in the identity of the enigmatic cyclist. "Recently, I have been thinking about the significance of circles in some research I am doing," McKane said. "I know this sounds weird. But I wonder what you think about the meaning of crop circles."

"I have been to them all," said the inscrutable figure. "They are all important to us, especially the ones near Stonehenge."

Without further comment or explanation, the wayfarer smiled, engaged the pedals of his bicycle, and rode away. He disappeared into the crowd streaming into the heart of Astoria.

FIGURES 3: Famous ringed crop-circle from Switzerland (Jabberocky). McKane adopted a similar image as the company logo, an oscillating electron. Below, a functioning Arc Reactor.

FIGURES 4: The fully functional Cold Nuclear Fusion Arc Reactor.

3

Paradox Meets Counter Paradox

Trinity: What is he doing?
Morpheus: He's beginning to believe.

The Matrix *(1999)*

As the biking oracle turned east, the father-son research team took a westward path. Packing up, they drove to their favorite golf course beside the peaceful Pacific. Their aim was to seek a brief distraction from covert research duties in their secret lab on the Northwest coast.

They felt they had earned it.

On the third hole beside the glassy Pacific, McKane bent down — not to pick up an errant golf ball — but to pick a flower. He raised the delicate white-petalled specimen to eye level, catching a glimpse of divine beauty encased in its ancient design. The tiny daisy seemed to remind him that *all things have universal significance, regardless of size*.

After a decade of work studying complex interactions at the level of an electron, McKane was beginning to believe in the interconnectedness of all events. He recalled the admonition of Paul Dirac, founder of quantum mechanics, who boldly defied classical ideas of time and space.

"Pick the flower on earth, and you move the farthest star."

McKane lodged the flower behind his right ear, respectfully honoring its place in a shared universe. As he took aim at the flag on the third hole, he realized that sport, like invention, was an unending quest for transcendent perfection.

* * *

On the fourth hole, McKane noticed a threesome on the green directly in front of him. The group included a man and woman in a gas-powered cart. The third player, a green cardigan-wearing gentleman of diminutive size, was twirling a golf club above his head, like a Majorette tossing her baton in a high school marching band. With each successful catch of the baton, the curious character energetically danced around the pin, performing an unmistakable Irish jig.

* * *

McKane and his long-retired psychologist father returned to the clubhouse after a satisfying, but imperfect, round of golf. Entering the foyer, they noted the presence of the couple from the golf cart and the dwarfish player with the twirling golf club.

The Irish dancer approached McKane.

"Oh my god, you are here — the guy with the flower in his hair. You are the one!"

Though mystified, McKane nevertheless nodded at him.

"Hey, I have a question for you," continued the stranger. "I am not gay or anything. But can I feel your hair?"

Not wanting to offend the unconventional golfer in a public space, McKane blurted, "Sure, if you want. I guess."

The stranger motioned for McKane to lower his head, as a king commands his subject to bow before him.

McKane obeyed.

The foreigner's fingers carefully combed through McKane's hair three times, like a geneticist collecting evidence for a DNA test. "Oh my god. It is such an honor to meet you," he said.

As the speechless inventor departed the clubhouse, a wild thought occurred to him.

What if someone in another space and time just picked the flower....

* * *

McKane took a long-anticipated phone call from the director of a high energy storage company rumored to have deep links to the government and the military. Also present on the Zoom call was McGregor, the company executive. Moose was the meeting facilitator.

McKane remembered their conversation well.

"Is this meeting being recorded?" McGregor asked the facilitator, conveying the need for both seriousness and secrecy.

"No, sir," replied Moose.

"Okay, that's good."

Turning his attention to McKane, McGregor stated his reason for the meeting. "It's a privilege to meet you. My staff briefed me on what you have been doing. I reviewed your patents. I am interested in talking because there are a fair number of applications for your inventions. They will link up well to other technologies. To what extent are you available to share your intellectual property?"

"I spent the last seven years in a garage developing a new form of energy. I don't have a job. I am unmarried and have no children. I just do research and write patents," McKane said.

"I can get you a job working on a Navy base that has no proximity to water," offered McGregor abruptly.

McKane understood the implications of the offer. He had heard the stories of a military facility that studied alien technologies.

"Okay, can you tell me more?"

"Everything is a system within systems. I see there are many places your science will fit in. Some of the locations are in classified spaces like the Navy facilities in Indiana. Other places are in national laboratories — Los Alamos, NASA, and so on. There are also a few private companies that might understand what you are doing. Your work could serve an important function."

McKane was dumbfounded. After years of derision and dismissive comments, someone with authority finally understood the implications of his work. He didn't know quite how to respond.

McGregor posed a leading question, "Does your technology work better in air or the vacuum of space?"

"Better in space, much better," McKane said.

McGregor paused and cleared his throat before continuing. "If you answer my next questions right, I can have you working with the government. I can arrange a phone call with forty-two NASA leaders at one time. This means you could share in solving some of the world's greatest scientific challenges. I could help you by making all the introductions."

"Okay," said McKane flatly.

McGregor turned to address the facilitator in a barely discernible whisper, presenting a procedural question. "Do we have NDAs signed all around?"

"No sir," replied Moose.

"And we are not recording, right?"

"That's correct."

"Good."

"Now McKane, I am going to show you something on my computer screen," McGregor said, opening an illustrative document. "I want you to look at the image closely. Tell me if you see a big problem here. And if so, can you tell me if there is a big solution?"

McKane studied the illustration as if it were an excerpt from an elementary school biology textbook. He recognized it immediately.

"That's easy. Your drawing is a representation of the carbon cycle. It shows atmospheric movements between the clouds, ocean, land, industry, and then back to the clouds. The immense blue image there — the ocean — is the big solution to the climate crisis."

"That is correct," McGregor stated point-blank. He abruptly closed the first image on his computer screen.

"Now, I am going to show you something else. It is a spreadsheet. On the top is a list of the things we know how to do. On the bottom is a list of things we don't know how to do, but we know are possible. We are attempting to learn the technologies on the latter list. Just read the words and tell me the first thing that pops into your mind. Just say the words."

McKane took a minute to review the rather complicated document. It was written like a scientific wish list. "OMG, it's the dendritic cells. That's the answer. I can do all types of dendrites in my garage, all organic and non-filterable. I can turn Costco bottled water with a pH of 7.5 into dendrite water with a pH of 3.0, the same as fresh white wine. The treated water has organic life in it."

McGregor solemnly closed his computer's screen-sharing function.

"We have been waiting for you over 35 years. There is a prophecy about you. Your impact on humankind will be considerable," McGregor announced in a surprisingly matter-of-fact way.

McKane went mute, stunned by the enormity of these words.

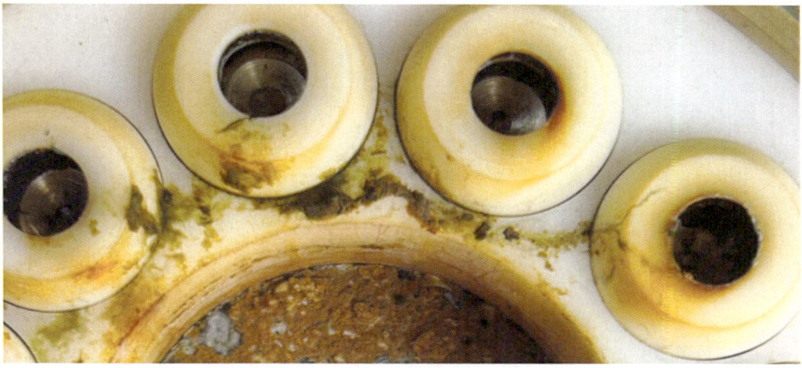

FIGURE 1: Circles-within-circles embedded in QKC's Arc Reactor. Organic (dendritic – polysaccharide – nucleic acid) sediment growth after 7-days with only 0.5 watts of electrical input. A creative force stimulated by the Quantum Kinetic Well.

"Well, now," McGregor continued, "Can you tell me how much money you have in the bank? Tell me if you have six million dollars."

"No, I don't."

"That is the amount of money you will need to initially defend against someone stealing your IP," McGregor explained.

"How many scientists would it take to back engineer your stuff and how long will it take?" McGregor asked. He spoke in a lower voice now, like someone keenly aware of an ominous threat close by.

"It will take four to six top engineers about four years or more to figure it out, if they can," McKane predicted.

"That is exactly what I was thinking," McGregor said.

Continuing with his line of direct, probing questions, the executive asked, "Did you put all your secrets into your patents?"

McKane noted that the integrity of the subject IP was important to McGregor.

"The very first thing my IP lawyer told me was to *tell it all* in the patents, or he couldn't fully protect me going forward. That's the rule of patent law as I understand it," McKane answered.

"Sure," McGregor said. "But remember, trade secrets are your best friends. You need to have several trade secrets embedded within your systems."

"I see that," McKane conceded.

McGregor gave him a stark warning. "People will try to steal your IP. If Samsung or some other multi-national corporation took your technology, there is nothing you could do about it, unless you already have one hundred million dollars in the bank. Some companies have a history of doing this. Since you don't have many arrows in your quiver, you will need to work with us or find a venture partner worth a billion."

Before McKane could respond, McGregor paused his soliloquy to confer with Moose in hushed tones. After a short, confidential back and forth, he returned to the dialogue at hand.

"The leaders of the top research organizations understand that innovation is a hard thing. Some will be excited about your work, perceiving you as headed to the top of the S curve. That's significant because you could share in solving some of NASA's greatest challenges."

McKane listened intently, as if engaged with an actual fortune-teller.

"Your technology will be very disruptive across many industries. You will encounter opponents who will attempt to decelerate the application of your inventions. They will try to lag the public's attention on you to delay access to the benefits of your intellectual property."

As if a silent alarm had suddenly been sounded, Moose unexpectedly interrupted the conference to remind McGregor of the hard break at the top of the hour.

"Sorry, McKane, but I've got to go," the executive said. "There is much more to say, but next time. It was a pleasure to meet you. Our conversation far exceeded my expectations."

* * *

The promised "next time" conference with the energy executive never came about. McKane was curious. *Why would McGregor drop out the promising dialogue after predicting an unknown inventor would appear from obscurity to alter the energy technology landscape?*

McKane reached out to Moose for an explanation and update on McGregor's intentions.

In a telephone call, Moose reluctantly reported that McGregor had recently started another research company focusing on a new technology. Moose then confided, "Sorry McKane. I can't explain further. Confidentially, I think you did it!"

4

Thunderheads on the Horizon

Tell me a story 'bout a fella with a crazy dream
Everybody tells him that it won't amount to anything
He fires the shot heard around the world
He wins the prize, and he gets the girl
That's the kind of story I like, the kind that makes you think

Steve Earle, "The Kind"

Leaving the lures of paradise gave McKane an empty feeling. As the Alaska Airlines jet reached V2 speed departing from Kawaii, his thoughts raced backwards in time to relive the final moments with Sommer — haunted by her questions.

What is this mysterious stone you gave me?

What are your intentions?

When will you become an islander?

* * *

Back on the mainland and alone in his secret laboratory, McKane reflected on the deep emotional toll his painstaking research cost him. He had taken the path of sacrifice, forfeiting a marriage, his coaching career, and a restorative relationship in Hawaii — all for the remote chance of achieving something unequivocal in physics.

McKane had imagined that the world would welcome novel research into the mysterious power of the electron. He thought an invention that efficiently harnessed vast amounts of energy from any water source couldn't help but be heralded as a trailblazing achievement.

But instead of earning acclaim as a climate change hero, he was now on the fast track to total obscurity. He and the business he had built. To be sure, some, like McGregor, had shared predictions that McKane's work could bring about something remarkable. But scant support from key voices and investors in the domain had materialized to substantiate McGregor's claim.

Scientists and technicians on the ground generally regarded his work with doubt and suspicion. The biggest investors in energy had yet to extend or even open their arms to McKane.

He was at the helm of a fledgling company, which indeed had in its possession remarkable technology. But without substantial partnerships — without powerful voices in the clean energy space at his back — there was little he could do to leverage his inventions into positive and practical change.

Despite the trail of disappointments behind him, he realized there could be no turning back from his chosen path. The odds were still stacked against him. But he had to push forward. Doing the hard work of an innovator. Making impossible discoveries. Establishing a framework for a radically new clean energy system. To McKane, these aims were not simply profitable ventures.

They were his *calling.*

* * *

The next target was scaling up the Quantum Kinetic Well, his crown jewel achievement, to the point of readiness for industry. McKane projected that the technology could lead the way to stop climate change, but only if properly placed in commercial space. His plan meant there was even harder work ahead outside the laboratory. He had to find ways to better educate the public, build up a team of dedicated supporters, and defend against privileged commercial competitors attempting to shun his disruptive technology from public view.

Knowing history was cyclical only reinforced his concerns about the trajectory of the business. He understood a terrible truth from studying the history of human progress — science itself had often been thwarted by discriminatory and sometimes vicious tactics, employed by powerbrokers and legacy institutions to maintain their establishment niches. Aspiring, but poorly funded inventors who dared to introduce disruptive thinking into mainstream systems were easy targets for defamation and banishment. Many discoveries and countless nascent technologies had ended in this way.

And the results only impoverished humanity and the planet.

McKane often reflected on the notable sagas of persecuted philosophers and forgotten inventors of the past. At the top of his list of the most persecuted thinkers of his day was Giordano Bruno, a sixteenth century teacher and ancient astronomer, who openly conjectured the universe was infinite and that other solar systems existed. Since his theories were detested by the Roman church, Bruno was arrested on the charge of heresy by the Roman Inquisition and subjected to an 8-year trial. The inevitable verdict sealed Bruno's fate — execution in the public marketplace of Rome's Campo de' Fiori. History never recorded Bruno's last words because he was muzzled with metal plates and led like a silent lamb to his personal funeral pyre — burned alive for the cardinal sin of exercising freedom of thought about the creation narrative and the structure of the universe.

McKane was also anguished by the epic tragedy of Nikola Tesla, a Croatian scientist and pioneering innovator, whose inventions inspired the world to believe in a brighter future aided by his cutting-edge discoveries. From the beginning to the end of his career, Tesla endeavored to resolve some of the world's greatest technological challenges including safe and maintainable electricity.

Despite his admirable aspiration to alter the course of history through applied science, Tesla's career was marked by achievements that fell just short of success. After years of trailblazing research with alternating current electricity, his accomplishments were attacked and overshadowed by Thomas Edison, whose direct current discoveries won the "Current Wars" in the early Twentieth Century. And later, after Tesla made milestone breakthroughs in radio communication, his achievements were hijacked by Guglielmo Marconi, who patented Tesla's inventions and won the Nobel Prize in Physics as the "creator" of the radio.

Mentally and emotionally shattered, Tesla incurred a chronic form of mental illness distinguished by severe OCD (obsessive compulsive disorder) and delusional disease. Untreated for his major mental illness, Tesla fell into abject poverty and became socially isolated, preferring the company of the pigeons in New York City rather than his one-time friends and scientific associates.

FIGURE 1: McKane always eager to explain scientific concepts. Pointing to new ways of perceiving the Universe.

McKane identified most with another shuttered-off genius, Oliver Heaviside. Heaviside was a self-taught mathematician who found unique solutions for making distortion-free telegraph lines in the early twentieth century. Despite a lifetime of profound insight in mathematics and unmatched ingenuity in electronics, Heaviside was barred from the affluent inner circles of the Royal Society and Great Society of Engineers. Instead, he lived the life of a social outcast on a pathetic pension, forced to live off the generosity of family.

Somehow, McKane felt a deep kinship with Heaviside. Perhaps it was because Heaviside, like McKane, was a former athlete and gymnast before channeling his energy into invention and an unceasing investigation of the impossible. Heaviside boldly introduced disruptive technologies and made courageous pronouncements challenging the status quo. He overcame defeatist rhetoric and propaganda with technical efficacy.

But he also suffered for it.

* * *

McKane needed to stack up financial resources to fund research into the many applications of his quantum reactor. Without public or private backing, his promising technology would be forgotten.

He reached out to Zach, an AI engineer who scouted technology for a large venture capital firm. McKane hoped the global investment giant could help launch his water reactor into commercial space. The AI engineer was intrigued enough by McKane's pitch to look at Quantum Kinetics' patented inventions, published research, and website.

But after going over the data, Zach expressed skepticism.

Your claim about changing heavy elements like uranium is nearly impossible to believe. Elements heavier than iron cost energy to fuse, making the fusion reaction energy negative. It's nigh unimaginable to me that your device is capable of fusing light atoms and then would just continue to fuse all the way to uranium.

The fact of the matter is that cold fusion is the holy grail of energy production. It takes a substantial leap of faith to believe it is possible. It takes an order of magnitude more faith to believe it would be stumbled upon accidentally in your electrolysis experiments. And it takes insanity to believe all of that without irrefutable evidence.

Notwithstanding his skepticism, Zach was quick to soften his harsh rhetoric in the interest of appearing a neutral arbiter.

Your self-educated background in physics may be an advantage. In fact, I think that your self-education is a good thing, as one with such a background would inevitably view physics through a different lens than the typical student who traverses a standard curriculum.

And he added a prediction.

Although I assess there is only a one percent chance that a new discovery has been made and a vanishingly low chance that cold fusion has been unlocked, I try to be as open-minded as possible. In that spirit, I am more than willing to have my mind changed.

But your extraordinary claims will require extraordinary evidence. For that, I would need to see either a publication of the experimental materials and procedures such that it could be independently reproduced in a third-party lab, and the accompanying third-party conclusion. This is because you need an irrefutable demonstration of a device that produces more energy than it consumes.

* * *

In 2007 a group of business professionals and environmental specialists created a commission to deal with the country's clean energy issues. Named the "CleanTech Alliance," the push behind the organization was to expedite funding for clean energy projects in the U.S., while also fostering supply networks internationally. Headquartered in Washington state, membership ranks extended across 17 states and 4 Canadian provinces. Over its history, the enterprise worked closely with many emerging companies committed to a green energy buildout, including several mega-funded nuclear fusion research companies in Washington state. Through CleanTech, these companies received generous support from interested billionaires and the U.S. government.

McKane took an interest in CleanTech after he noticed it was actively promoting the big fusion companies yet never reached out to Quantum Kinetics Corporation, an aspiring clean energy upstart in Washington's own backyard. Despite the bureaucratic backhand to McKane's company, he nevertheless decided to join CleanTech early in 2023. He couldn't deny that it was a good business decision to open a door to connections that might pay off later, providing a platform for collaboration with other industries.

They'll come around, he thought.

* * *

It was conference day, June 22, 2023. Hundreds of members would meet in the Seattle Hyatt at Olive 8 for a CleanTech innovative showcase between 8:00 a.m. and 5:00 p.m.

Billed as an assemblage of "forward-thinking clean tech decision-makers leading the decarbonization change," those in attendance would include an impressive list of government leaders, industry executives, educators, tech scouts, and representatives of fledging startup companies eager to explore good markets with their budding energy solutions.

Also present were two members of the Quantum Kinetic team whom none of the qualified people present at the conference knew from Adam. For McKane and his father, Riley, it felt as if the two of them had just arrived from another planet.

The show began with a buffet and continental breakfast. McKane and Riley had showed up early, not taking their chances with commuter traffic jams that might make them late. The hotel entrance hall outside the banquet room was already jam-packed with conference attendees when they arrived. The two gentlemen from QKC, dressed in dark gray dinner jackets, white collared shirts, blue jeans, and tennis shoes, blended into the group—but not to mingle with acquaintances. Their eyes were set on the delicious breakfast of hot coffee, scrambled eggs, bacon, and oatmeal available to guests.

They took their delectables with them into a vast, empty auditorium that could easily seat four hundred.

* * *

With the ballroom now brimming to capacity, the conference moderator introduced the morning keynote speaker, Dr. Vanessa Chan, from the U.S. Department of Energy. Dr. Chan's position with the government was "Chief Commercialization Officer and Offices Director of Technology Transitions." A charismatic person and dynamic public speaker, she gave a rousing pep talk to rally the troops and focus attention on the importance of "believing in new technology for the world."

"We have to find better ways to do everything because we are running out of time," she said with a note of gravity in her voice.

After Dr. Chan's keynote address, the moderator asked for questions from conference attendees. The respectful audience fell into total silence, as if a hidden hand had pushed the auditorium's mute button. The hushed gallery waited for what seemed like minutes before an intrepid questioner raised a hand. His willingness to break the silence brought collective giggles of relief from the audience.

McKane took the microphone handed to him.

"Good morning, Dr. Chan. It is an honor to speak to you. What do you think about metal hydrides as an innovative technology for the future? And if you had to pick a specific metal hydride, which one would it be?"

"I really don't want to talk about metal hydrides right now, but we can talk off-line," Dr. Chan responded. "As for your second question, it's just like Pokémon. *I gotta catch 'em all.*"

* * *

At the morning break, the restless throng of scientists and entrepreneurs moved in unison to the entrance hallway where fresh coffee and pastries were being served.

McKane, however, broke off from the crowd and headed for an anteroom off the corridor. He opened a closed door and entered another large chamber resembling a concert hall.

Standing on the opposite side of the room was a singular figure filling her pink canister with fresh water. It was Dr. Chan.

McKane approached her.

The brilliant scientist did not run away or call for security. Instead, she seemed to anticipate another question from McKane.

"I am the guy you wanted to talk to about hydrides, but not in front of everyone," McKane said.

"Oh, yeah, that's right," she said.

"Check this out," McKane said. Pulling out his cell phone and opening his photo app, he scrolled through a montage of images from his laboratory as she looked on. Together they carefully examined photos that were new and had not been seen before by others. Pictures of organic, dark black, bright white, light yellow, and deep red hydrides arranged in stacked columns, grown from ocean water.

FIGURE 2: Fabricated oceanic dried carbon (organic) hydride (left), post excited hydride (middle), thermal excited hydride emission of clean-burning hydrogen flame (right).

Dr. Chan leaned closer to McKane, shoulder to shoulder. “I have heard of people like you. They claim to do what you do, but we have never been able to get the numbers. Can you get me your numbers? Is there a way for me to contact you?”

McKane handed her his business card. She flipped it over several times.

“I really like your business card,” she said. “It is very mysterious.”

“It comes from a crop circle,” McKane admitted. “It’s actually a representation of an oscillating electron.”

“Okay. Well, just get me your numbers. Here’s how to contact me,” she said, handing over her card.

“Thanks. I will get those numbers for you, but can you finish the story?” asked McKane.

“What story?”

“You said something in your speech this morning that resonated with me. You said we are running out of time. How much time do we have to find the secret before it’s too late?”

“Three years,” said the famed MIT physicist, as if reminding herself of a foreboding storm on the horizon. “We only have three years.”

* * *

Both returned separately to the conference hall, just as new speakers mounted the stage to join the lunchtime keynote panel. The topic was securing minerals from the earth’s surface “to aid in decarbonization of the environment.”

While hotel servers delivered delicious chicken salads to the spectators, Dr. Aaron Feaver, the Director of The Joint Center for Deployment and Research in Earth Abundant Materials (JCDREAM), led off the presentation by describing recent efforts to establish sustainable material supply chains "for clean energy and transportation." Dr. Feaver highlighted the need for silicone, carbon, lithium, platinum, and iridium to support "zero-emission technologies" in the U.S. and to credibly compete with China—a geographically vast country with enormous mineral resources.

Another panel member, Rick Luebbe, CEO and co-founder of a battery materials company, echoed the need for silicone batteries "to power the electrification of everything." Both Mr. Luebbe and Dr. Feaver joined to report on a new mining project in Idaho, an exciting opportunity for their business ventures because Washington State had forbidden their environmentally harmful mining practices. The lunchtime presentation wrapped up with an array of pictures of massive machines and barren, devastated landscapes in Idaho — coming soon to Washington State, provided it were politically possible.

FIGURE 3: McKane at the CleanTech Alliance conference ready to engage with the question-and-answer section.

Facing an audience of green energy enthusiasts, Jennifer, the panel moderator, courageously opened the floor for questions and comments.

As before, there was silence and a growing tension in the great assembly. After what seemed like an age, a familiar young man raised his hand again.

Laughter and applause resounded through the great hall when McKane took the microphone once more. Their response was evocative of ancient crowds cheering on a gladiator as he stepped into the coliseum to face an opponent.

McKane's question cut straight to the heart of the matter.

"Did you know there are 352 quintillion gallons of water in the oceans? And within that measure, there are 50 quadrillion tons of rare minerals dissolved in the ocean. So why are you willing to destroy swaths of the earth's surface in the search of rare minerals for battery production when we are surrounded by a big blue solution?"

With murmurs and chuckles ringing at the unfolding spectacle, Dr. Feaver feigned and parried. "Don't you know seawater is too dilute to be useful?"

Quickly, Rich Luebbe stepped between the combatants and said, "Okay, I want to jump in here and give McKane an opportunity to tell us the answer. Just tell us. You seem to know. Now we want to know."

McKane looked at Jennifer, as if seeking permission to tell a dangerous secret. Jennifer shook her head. "I'm shutting this down now. Next question please."

At Jennifer's signal, her assistant snatched the microphone out of McKane's hand and moved on.

Disarmed, the young warrior collected his gear and exited the arena.

He soon noticed that he had been followed closely by a cadre of guests also on their way out. They met him in the banquet room hallway where an impassioned conversation struck up about what had just transpired in the conference room.

"OMG, that's exactly what I was thinking, but I didn't stand up and say it," said one follower.

"Who are you, anyway? Do you have a business card?" said another attendee.

"Hey, you are going to need a resource guy to manage all your money. Here is my card. Give me a call," said a third conventioneer.

After responding to each inquiry, McKane moved down the corridor towards the entrance of the Seattle Hyatt at Olive 8, intending to leave. He suddenly felt the presence of others nearby. He glanced behind to see other members of CleanTech leaving the conference just behind him.

McKane watched them for a minute, curious about the movement he had apparently started.

So, this is how it begins, he thought.

* * *

Before the next meeting with the CleanTech Alliance, the undaunted inventor reserved a large conference room at a posh hotel to be ready for the arrival of officials led by Tammy Deets, an international trade specialist. They were to be accompanied by Liz Laxague, the Global Environmental Technologies Team Leader of the US Department of Commerce and other governmental dignitaries, who had been assisting U.S. companies for many years. A representative of the governor's office would also be in attendance.

McKane wondered how the diverse group would get along. With a slight chill, he realized the collective reaction to his controversial Power Point presentation could easily backfire.

* * *

McKane surveyed the assemblage around the expansive conference table, including the tireless and effervescent Tammy Deets. He was disappointed to find that Liz Laxague from the federal government was not present, having canceled at the last moment for an unstated reason.

The official from Jay Enslee's office was late by 10 minutes. When he finally arrived, McKane noticed the stern grimace on the man's face. He wore the demeanor of a battle-ready government recruit pressed into hard service.

Donuts and coffee had already been served. Sounds of welcome and congenial small talk echoed in the chamber, signifying affable friends who were glad to see one another. McKane sensed that this conference was not his guests' first rodeo together. In fact, to them he was probably a dubious newcomer to the old guard.

McKane called them to order and started his presentation. It was a PowerPoint slide deck that would have persuaded any open-minded jury. McKane described his six years of solitary research in a garage. He recounted a momentous discovery that came about fortuitously and an additional two years of confirming research in a joint strategic partnership with the Department of Energy. He reported that their research had been published in a *Nature* journal article with worldwide distribution. He generously provided detailed descriptions of the operative effects of the novel reactor.

After summing up the many commercial applications, he offered time for the attendees to ask questions.

Tammy was the first to react to the presentation.

"Okay, thanks for the information about your clean technology," Tammy said. "My question might sound out of left field or a little weird but hear me out. As you described how your reactor works, it occurred to me you were describing a blackbody in space. I am not an astrophysicist, but I wonder if your invention is some sort of micro-sized black hole?"

"Intriguing and good question," McKane said, being careful not to criticize an intuitive inquirer. He realized the scientific community could benefit from the same approach. *Ask the big question first.*

"Are you suggesting, Tammy, that it might be possible to build a synthetic un-ignited star on earth as a clean energy generator?" he said.

"Not exactly, but I suppose that's a better question," she said.

"Those are your words not mine, but I appreciate your thought-provoking question," McKane said. He was speaking with uncharacteristic caution, trying to hold the interest of the skeptical in the room.

As if a fire alarm had just sounded in the conference room, the Washington State official from the governor's office loudly cleared his throat. Having interrupted the flow of conversation, he then addressed the room like a professor trying to get his students' attention back on topic.

"Hey guys, I really like my job and my new office with a great view of the water, so please stop. I don't want you to go any further."

Having unilaterally kiboshed the meeting, the state bureaucrat shot dismissive words at McKane. "Hey man, I was in the Navy, and we don't talk about such matters publicly. We all believe great things in clean energy are possible. Fusion is possible. A lot of very important people have already put their chips in on fusion in Washington State. We are working with Commonwealth, Avalanche, Zap Energy, and Helion Energy — the people doing fusion in Washington. I am out of here and back to my office."

With that, the state representative stood up unceremoniously, grabbed a snack off the table, and on his way out squeezed McKane's hand so hard that McKane felt like his fingers might explode.

Before he left, though, the official made a point to pull McKane towards him until their chests were touching — too close for comfort.

"Welcome to the valley of corporate death," warned the state official.

* * *

Later, Tammy reached out to extend apologies for the adolescent bullying at the conference table. "I am so sorry, that didn't go as I hoped." She again offered her assistance to the fledging company.

"It was great to talk to you. Thank you very much for your information about Quantum Kinetics. Based on your business plan, I recommend you expand internationally. The World Nuclear Exhibition is a fantastic trade show, and they have a startup program that may fit your needs. The show is in France, and it is a good market for nuclear solutions. I would love it if you could accompany the Washington state delegation this year."

5

You Grow Uranium!

Everything is theoretically impossible, until it is done.
Robert A. Heinlein

Breaking through the noise of public criticism and McKane's growing self-doubt, an important message arrived from Ed at the national lab in Colorado.

"I looked at your before and after treatment specimens from the Columbia River water. There is a confirmed presence of Uranium-234 and Uranium-238 in your samples. I don't really know what you are trying to do, but you grew Uranium by 50% in twelve hours. Quite frankly, surprising …"

* * *

After three days of field testing at his makeshift lab near Astoria, McKane reached out to Nick at the Pacific Northwest National Laboratory (PNNL) in Washington State. He hoped Nick could steer him out of his lonesome laboratory and into a world class facility that did radioactive water assessments. Nick was happy to hear from his old research partner.

FIGURE 1: The post experiment electrodes from Hanford River water treatment 30v (left), 60v (middle), and 90v (right) from the *Quantum Kinetic Fusor.*

Together, they had previously conducted 18 months of experiments with the Quantum Kinetic Well and then co-authored a paper featuring the discovery in a *Nature* journal. McKane wrote the first draft, including an overview of the apparatus and the specific experimental design. Nick pitched in by describing the technical analysis of the pre- and post-treatment samples. The article had been accepted and published by *Nature* in record time. The three esteemed *Nature* reviewers heralded the device as "a much-needed invention around the world."

<u>FIGURE 2:</u> View inside the active reactor known as the "*Quantum Kinetic Fusor.*" Prior research with the DOE utilized this style of isotope manipulation of well water and ocean water. Findings were peer-reviewed and published in npjNature: Clean Water in 2022.

It had been months since McKane and Nick had spoken. It was good to catch up, including updates on salmon fishing in the ocean, their shared passion outside of the laboratory. The Washington Department of

Fish and Wildlife was already on record — the summer season would be exceptional.

Nick was eager to help his buddy with the continuing scientific inquiry into the patented water-based energy reactor. Regrettably, PNNL was not set up for the testing of radioactive isotopes. McKane would have to look elsewhere. Nick knew of another national lab that could do the job. It was in Colorado.

"Here is the contact information. Give them a call. You can use my name and reference our work together at the national laboratory in Washington state. They will be happy to help you." Nick said.

* * *

In June of 2023, after several ghosted email messages and dead-end phone calls to the Colorado lab, McKane somehow sweet talked his way past the receptionist and into the inner sanctum of facility decision makers. Ed was the research team leader.

McKane reached out to Ed. The inventor argued for the critical need to assess a recent test treating Columbia River water with his quantum reactor. He explained that the patented and research-verified invention could change the atomic identity of radioactive isotopes, reducing these "bad guys" to the point they could be stored safely underground or used for commercial purposes.

It was an important issue in Washington state. After all, Hanford was known to be the most radioactively contaminated place on earth. The magnitude of Hanford's threat was geographically immense, endangering the Columbia River from Eastern Washington to the Pacific Ocean 230 miles downstream.

Lives and livelihoods were at stake, McKane contested. *An entire ecosystem is in danger. There presently is no viable plan to destroy used nuclear fuel. Without a plan, the threat will exist for millions or billions of years.*

McKane posited that his machine could provide an effective method to alter levels of radioactive waste by synthetically converting dangerous nuclides into inert elements — something previously thought impossible in physics. He further explained that the water-reactor's cleaning effect was triggered by the consumption of hydrogen electrons at low-injected power and at room temperatures. The method worked with ground water, ocean water, distilled water, river water, or even tap water with no electrolyte added.

He disclosed his vision of cost-effective, retrofitted decontamination facilities at existing nuclear plants on the Columbia River, in the US, and beyond. Moreover, this technology could universally reduce environmental concerns about nuclear power, leading directly to the expansion of nuclear power facilities and the concomitant reduction of global greenhouse gas emissions.

McKane called it "Safe Nuclear."

Ed was initially hesitant. Despite the allure of the young man's daring claims, he was looking into the face of a previously unknown inventor, someone without proper academic credentials. Nevertheless, he was intrigued by McKane's ardent advocacy and fascinating scientific propositions.

Ed considered the odds. Despite the improbability that McKane's test of highly diluted river water would reveal significant isotopic changes, the inventor's hypothesis about transmutation of nuclear waste might be right. If so, this technology could be a game-changer in the field of radioactive waste remediation.

Given the stakes, Ed agreed to weigh in on the question with an open mind.

McKane's persistence and persuasiveness had paid off. A renowned researcher was willing to jump into the breach to facilitate the analysis of what was admittedly a phenomenological long shot. McKane was elated with the team leader's decision. He wasted no time asking for instructions on how to transport the samples from the Washington coast to the Colorado lab.

* * *

Two months passed without a word.

Then McKane received the long-awaited phone call from Colorado. It was Ed, wanting to give a heads-up on the test results. The lab had completed analysis of pre- and post-treatment Columbia River water samples. His report enthralled McKane.

> *I looked at your samples from the Columbia River. You need to know that the river water samples were fairly diluted as to the radioactive elements. But your treatment process increased Uranium significantly. I don't really know what you are trying to do, but you grew Uranium by 50% in twelve hours. I don't know how you did that. This finding is remarkable and demands further investigation.*

Additionally, Ed offered an insightful hypothesis.

> *Since you can substantially increase Uranium, I am wondering whether you can go the other direction and reduce it—thus doing a rapid radioactive cleanup. We should contact a lab in Georgia that can work with more intense radioactive solutions. I will contact them and set it up.*

* * *

The titanic meaning implicit in Ed's evaluation of the preliminary evidence was not lost on McKane. A national expert in isotopic science assessed it possible that an unknown and potentially miraculous phenomena was more than the product of a young man's overactive imagination.

More importantly, his report finally debunked the claim that McKane's ideas represented a whimsical notion belonging to the realm of science fiction rather than physics.

In reality, McKane's bold research was on the precipice of proving sustained nuclear fusion for the first time in history. McKane had coined this process *"Fast Fusion,"* because it edited isotopes within hours or days rather than thousands or millions of years. The potential benefits of *Fast Fusion* to the world were unimaginable. With technology like this at hand, humanity could control its destiny on this planet and beyond.

6

Into the Time Warp

There are really four dimensions, three which we call
the three planes of Space,
and a fourth, Time.

H.G. Wells, The Time Machine

On Tuesday, June 20, 2023, Ed wrote to the Georgia lab.

I am contacting you about a project for McKane Lee and his company (Quantum Kinetics Corporation). This is not my lab's project, per se, but McKane and his company were using us for analysis of before and after treatment samples. The initial pre and post treatment samples we were sent did not have a significant amount of any of the isotopes present. You could say the uranium isotopes (U-234 and U-238) were "significant" in that they were above the MDC, but only at about 1 pCi/L.

I believe he took those values from some multiplicative factor above the MDC we achieved for some analyses we recently performed for him. The actual gamma isotope activity is not "critical"–McKane just needs to know how much is there before treatment and after treatment so that they can determine the effectiveness of the treatment. He and I had a phone conversation, and the thought was that if they wanted to ensure a 99% removal of an isotope, they may want to have 100x the amount present compared to the MDC achieved. This is a rough ballpark, and I don't know what multiplicative factor or even MDC for each isotope he used.

I would question the need to have a 97% enriched uranium line there. It would have to be its own sample (separate from the natural U). In either case, uranium is uranium—I would expect them to behave the same "chemically" and "physically". As you may know, to separate the isotopes to enrich the U-235 is a significant effort in special circumstances, and even then, it is hard to separate it from the U-234 (U-238 is reduced in mass, but U-234 remains with the U-235).

McKane also reached out to the Georgia laboratory.

Quantum Kinetics Corporation just needs to "dirty up the water water" (no pun intended) a little bit more to get a clearer picture of what QKC's patented device is doing to contaminated water. We already know from two years of prior testing with the Department of Energy that it is manipulating smaller isotopes including the hydrogen and oxygen isotopes and producing carbon-based magnetite crystals, but QKC needs to explore the full potential. From the last experiment, though the values were low yet barely detected, Uranium-234 and Uranium-238 went up post treatment.

QKC wants to ensure all parties (both labs) are happy with the parameters of this "unique" experiment. And I appreciate everyone's efforts in this.

Again, thank you both for working together on this brain-busting exercise. Very much looking forward to what we can accomplish and learn.

Christina, a PhD at the Georgia facility, responded and agreed to coordinate the testing setup.

We have been discussing this with our production team. Just FYI, our typical low-level tracers are on the order of 0.1 – 1 Bq/mL of solution, o~10 nCi/L. Your requested activity concentrations are 10 to 1000 times lower. Would you be willing to accept a nominal activity solution, knowing that any individual containers all came from the same single 4 L batch and share the same activity concentration? To make a solution like this, we would need to start from calibrated solutions and do gravimetric transfers. At such low activities, the initial solution made from gravimetric transfer would have to undergo serial dilutions, which would introduce additional uncertainty after each dilution. I understand what your targets are: Cs-137, Co-60, Cr-51, Co-57, Sr-90, Am-241, Pu-240 (primarily), and Uranium (natural). Our team is still discussing this request.

***

In the wake of the announcement by the national lab in Georgia — its agreement to evaluate an unprecedented process which the inventor maintained opened the envelope of time — McKane wanted to ensure all the details of transporting the modular reactor to Georgia were properly in place. Once the experiment date had been set, the laboratory space reserved, the lab materials readied, and sunk costs incurred there could be no turning back.

The young researcher felt the strain of the challenging task. No one had ever attempted to move a fully functional nuclear fusion reactor coast to coast on a commercial aircraft. To be sure, there were aviation regulatory rules to be met and national safety standards to be observed. The aim of the Quantum Kinetics team was to follow all governmental direction and guidance.

According to plan, McKane gave a heads up to the Transportation Security Administration (TSA) of the Company's need to transport the modular electronic reactor from Washington state to Georgia where a national lab would put the device under a strict peer review process. He first reached out with messages to the TSA by email and then by phone. He asked for directions on how to ensure there would be no administrative issues in transporting the specialized scientific equipment.

The return comments from the federal officials were easy to understand — the device could be considered a "permitted item" as carry-on or checked baggage but only after a special screening process at the airport. This meant the final decision would be up to the TSA officer after physical inspection.

* * *

McKane and Dreke walked through the doors of SeaTac International airport three hours before their scheduled flight. Dreke, a retired Boeing executive, was McKane's unofficial chaperon for this trip. They had showed up early, not taking their chances with unexpected inspection hiccups or unknown mandates that might make them late.

In the hands of the two researchers were two carry-on pelican cases. In the foam padded travel compartments were the segments of a safe fusion reactor. Hidden in the Quantum Kinetic Fusors were solutions to some of the world's greatest problems.

<u>FIGURE 1:</u> The two pelican cases transporting the *Quantum Kinetic Fusors* and associated lab equipment through SeaTac.

* * *

McKane placed the pelican cases on a conveyer belt moving into the X-ray machine. He took off his wristwatch, shoes, belt, and jacket — minimum requirements before entering the surveillance chamber. After successfully reaching airside, he turned to see if the Company's carry-on baggage had also cleared the security checkpoint.

McKane watched as the transportation security officer studied the odd images displayed on a radiographic screen. The official carefully panned, scrolled, and rotated the pictures in 3D-space. What she saw was a device resembling a prop from a *Star Trek* movie, the kind with the power to bring life to dead planets. She leaned forward to get a better look at the wonderous object, mystified or in shock.

Suddenly, the conveyor belt stopped.

Abruptly halting the inspection process, the specialist swiveled in her chair and called out to a Federal Air Marshall (FAM) positioned nearby.

The Marshall darted over to the X-ray station as if she were responding to an emergency signaled by a siren's blast. Leaning over the inspector's shoulder, the FAM appeared more amazed than concerned. She stared at the viewing screen. Then she pointed to the monitor and spoke to the inspector.

McKane didn't intend to eavesdrop, but he couldn't help but hear the words "OMG, this is interesting. What is this thing?"

After a drawn-out examination of the device that seemed like an eternity to McKane, the FAM stopped grilling the inspector, shrugged her shoulders, and gave a firm hand signal to roll the conveyor forward.

* * *

McKane was nervous when he boarded the jet plane departing from Seattle, set to arrive in Georgia in five hours. Dreke was excited to tag along for the ride. On this trip, he would be present to see something other than a plane get off the ground and flying — that is, if everything went as McKane hoped.

McKane admitted his frayed nerves to Dreke. So much was at stake in this upcoming test, which involved high intensity spiked radioactive cocktails. No one had ever tried to synthetically alter the universal time clock before. He recognized that he and his companion might be witness to a historic achievement — or else a total flop. Dreke assuaged McKane's rattled nerves by reassuring the inventor that his reactor would test out flawlessly.

"We got this, buddy," Dreke said.

* * *

On the morning of the test, the facility safety manager, two research officials, McKane, and Dreke met in the laboratory's conference room. The safety officer handed the team members their respective dose meters. The buzz of excitement in the air about the prospects of a potential breakthrough in physics was palpable.

But the attendees' first purpose was to establish agreed upon technical procedures and ministerial ground rules for the upcoming experiment. In the next 24 hours there could be no surprises. Nothing to invalidate the results. These ground rules had to be executed flawlessly for the outcome to earn the public's respect. In short, their experiment would have to meet the most rigorous scientific standards.

"Once you do certain things with the radioactive elements, you can't go back, " Mary said. She was the lead researcher, a brilliant woman with three letters behind her name. It was clear she would control every step before, during, and after the test, including performing the critical task of mixing the toxic brew. She would also marshal the test area, deciding who could enter the lab once the reactor was engaged and operational. *Safety first* was the plan. No one dared to question her authority. After all, she had an impressive professional background, including operation of one of the largest particle accelerators on earth.

Before starting the test, the lab officials wanted a quick look at the quantum reactor to gauge the size of required shielding and venting. As

McKane unboxed the invention, the atmosphere in the room was charged with nervous anticipation, like a family in a maternity waiting room anxious to see a newborn for the first time.

Rick, the facility administrator, could barely disguise his enthusiasm. "It's so great to work with you, McKane. We have all been looking forward to meeting you and helping you with this experiment. You come with good recommendation from Ed, a guy we really respect."

Tina, the research assistant, chimed in to say that everything was properly prepared for the test. All the radiation safeguards were in place.

Mary added, "Given its size, the device can be easily shielded. We can use more shielding around the equipment as necessary."

The Georgia lab officials wanted to start early in the morning. There would be three hours allotted for prep time before "time zero," which was 11:00 am. This meant 8:00 a.m. was everyone's start time. Mary would prepare the 4 ml solutions immediately before the test, pouring the radioactive substances into containers with leak tight gaskets. Protocol called for the use of Millipore, i.e., ultra-pure water. Mary had 100 liters set aside. There were ample backup supplies if more was needed.

Mary insisted that McKane should be involved at every step. "Just be sure you don't touch the inside of the reactor," she warned.

The stream of supportive and respectful comments from the extremely competent scientists gathered around him lifted McKane's spirits. With a cloudburst of inner joy, he realized he was no longer a solitary inventor and wandering outsider, trotting this long, exhausting journey alone. He was now part of a larger community of scientific explorers pursuing discoveries that could ignite a new age of technology.

* * *

It was time zero. Present in the test area were the lab officials, along with Dreke and McKane. Mary, the quarterback and mastermind of the operation, took control on minute one, insuring proper execution of the experimental protocols. Tina assisted in the safety department by keeping a close eye on the electronic monitors. Rick served as technical observer and peer reviewer. Mary told Dreke he could also be in the test area because of his engineering background and career work in the aerospace industry.

Mary wanted final words with the young inventor before *go time*.

“Can you reassure me that the reactor’s operational power level will not strain our electrical capacity?” she asked in a commanding tone. “High voltage is more worrisome to me than high radiation levels.”

“The voltage is less than household levels,” McKane calmly replied.

“Okay. Then we don’t have a problem.”

To further reassure her, the inventor explained, “We call the process *Electro-Physical Transmutation* or *E-PT* for short. It transmutes atomic elements at low input power. Today, we want to demonstrate the downshifting of radioactive elements into inert substances like iron. Water is the primary energy generator for the reactor, not external electrical power.”

“This is mind blowing stuff,” Mary said. “It appears your device could induce beta decay. If we get confirmation, this is a Nobel Prize for you!”

* * *

McKane stepped forward to inspect and ensure the proper set up of the quantum reactor. When finished, he turned to the attendees. “Okay, everyone come here. I am holding a voltmeter and signal generator. We are not powering up yet, but I want to show you something.”

FIGURE 2: The two *Quantum Kinetic Fusors* reactors setup within the fume hood ready to begin the experiment. Quantum Kinetic Fusor (left) set to 45v, and Quantum Kinetic Fusor (right) set to 90v input. The XR-100CR X-ray detector positioned underneath the 90v reactor.

The young innovator got on his hands and knees.

"Watch the needle here. In a few moments, you will see the voltage needle climb, even though we are not drawing more power from the wall. We are just changing the frequency."

"Oh, it's like a radio," Rick said.

"Correct. It's like a radio tuning into a broadcast present in the air. But instead of air, our device is locking onto the water."

They all watched the reading closely.

"Now watch what happens when I scroll through the frequency levels. I am not adjusting the power, but the voltage starts to fall. We find the right frequency by going past it and then back."

Dreke said, "Why is voltage so important?"

"Because the higher the voltage, the easier it is to crack the water, like sound waves shattering a wine glass."

"So, it's like what we do on jet wings to reduce resonance, to keep them from flying apart," added Dreke.

"Correct," replied the inventor. "The difference is that you kill the wing resonance while we use the resonance to split the water."

Continuing, McKane said, "Now that we have resonance established with the signal generator, we can attach the main pulsing circuit driver to the reactor's power supply. When we reach the kilohertz frequency matching the resonance of the water, the X-ray detector will start beeping as we increase the voltage."

"Okay. I am going to adjust the power supply to 90 volts. Everyone should step away now. We are going up to 150 million degrees Celsius. Is everyone ready for this?"

A hush fell over the room. No one said a word.

With no dissent from the other experimenters, the inventor stepped back and triggered the reactor with the relay remote switch. The X-ray detector started beeping wildly and the photon spectrum reached 15 keV – 150 million degrees Celsius within 2 seconds.

FIGURE 3: McKane turning to the laboratory crew to explain the power of the Quantum Kinetic Fusors.

McKane then turned to address the spellbound observers. He pulled his iPhone from his lab coat pocket. "Okay. You see this iPhone. It takes 20 to 40 watts to charge overnight. Look at the wall meter over here. How much power is the reactor drawing right now? It is only pulling 29.4 watts from the wall. This means we can ignite a safe nuclear fusion reaction with less power than needed to charge your cell phone."

Tina's face suddenly turned white. Wordlessly, she turned and hustled out of the test area and down the hallway.

"Your reactor glows like the Sun!" said Mary, astonished.

FIGURE 4: The power wall meter showing *Quantum Kinetic Fusor* reactors' (45v and 90v) total energy consumption of 29.4 watts, while sustaining nuclear fusion (15 keV – 150 million degrees Celsius).

* * *

Outside the test area, the researchers shed their protective clothing and gloves while they watched the reactor through the lab window. It was running perfectly on its own. They moved down the hallway and assembled in the break room for their second cups of morning coffee. Suddenly, Tina ran past the break room window and entered the test area.

"What is she doing?" asked McKane.

"She has a detector in her hand," Rick answered. "She wants to make sure no one is going to die."

Minutes passed before Tina exited the test area. She walked slowly by the break room window, making eye contact with the others. She gave the team an affirmative nod and an enthusiastic thumbs up.

* * *

Later, Tina found McKane as he was leaving for a late lunch. She had a message for the inventor. Leaning in close, she whispered, "I want you to know that you can use my name whenever you want. You can also reference me in your published research, if you like."

* * *

The following morning, McKane and Dreke entered the front door of the national laboratory, intending to head for the observation window outside the test area. They would wait there until the final whistle — exactly 24 hours after go time.

Before reaching their destination, a tall gentleman dressed in a collared golf shirt and khaki cut-offs called out, "Hey, are you McKane?"

Without waiting for a response, the newcomer approached and introduced himself. "I am Harrison, the globalist for the Company. I need to talk to you. Can you talk now?"

McKane looked at his wristwatch to check the time. "Absolutely," McKane said. "But we need to shut off the machine in 20 minutes."

"Let's meet in the conference room." Harrison led the way.

* * *

They sat down at a table with room enough to spread out briefcases and travel bags. Harrison got right to the point.

"Tell me everything you know about what's going on in the back room. I need to know."

McKane held his breath, momentarily taken off guard by the unexpected challenge. He glanced over to Dreke, who nodded back, offering McKane tacit approval to provide full disclosure of their work.

"Okay," McKane said as he walked over to a large white board. He grabbed three colored markers and began drawing.

McKane turned to address Harrison. "This is what we know about the quantum reactor," he said while jotting a list on the board, each written item accompanied by an illustrative diagram. "And this is what we don't know about the reactor." He completed the list and the expository drawings.

Finished, McKane set the markers down.

"OMG, it's a time machine!" exclaimed Harrison.

"Those are your words, not mine. However, I am not disagreeing with you."

Harrison made another demand, "Take me back there right now. I want to see your set up and the radiation readings."

* * *

After observing the Quantum Kinetic Fusor in action, Harrison said, "You need to do me a favor. I need all your metal ICP values and other results to take upstairs. We need to know if any new metals are forming in this process. I want you to know there are less than one percent of the people in the world that could appreciate what you are doing in here. And our company is part of that one percent."

7

First Dawn of Safe Nuclear

"The explorer who will not come back or send back his ships to tell his tale is not an explorer, only an adventurer; and his sons are born in exile."

Ursula K. Le Guin, The Dispossessed: An Ambiguous Utopia

Following the successful experiment in Georgia and armed with the confirming results from the independent lab in Colorado, the QK Team began work on a white paper to summarize its latest findings. In a world fractured by the stunning failures of politicians and industry leaders to find permanent solutions to the Earth's greatest problems, QK's researchers hoped publication of the manuscript would help rekindle civilization's aspiration and action towards a brighter energy future.

The experiment itself had been momentous. It involved the meticulous task of conjuring an exotic, heavily spiked radioactive cocktail — one that could kill any organic life form on Earth — and then running the toxic brew through QK's revolutionary process dubbed "Electro-Physical Transmutation (E-PT)."

What the test showed was that Quantum Kinetics had pulled off the miracle of rapidly accelerating radioactive beta decay at 2.40 — 29.4

watts of electrical input and at room temperature. Specifically, E-PT's epoch fusion power substantially reduced ^{137}Cs (Cesium), ^{90}Sr (Strontium), and ^{235}U (Uranium) in only twenty-four hours.

In contrast, Mother Nature's radioactive beta decay spans millennia to bring about the same isotopic downshifting. For example, ^{235}U has a half-life of 700 million years while a single-chambered QKW™ significantly reduces ^{235}U in a matter of hours. The invention also alters ^{239}Pu, $^{234/238}U$, ^{241}Am, $^{57/60}Co$, and tritium. To further enhance the E-PT effect, simply stack the modular designed chambers in series or parallel configuration, to propel the process not in a linear rate, but an exponential rate.

Complimenting the triumph, the study revealed E-PT consistently transmuted radioactive elements into useful elements such as iron, silver, Tungsten, Thorium, Magnesium, and Barium. The achievement suggested QKC's isotope editing power was tantamount to modern alchemy — the fantastical and terrifying power to alter atomic elements with the turn of a simple electronic dial.

* * *

Believing the good news of the experiment could only be propelled by strategic media outreach, the QK team determined to launch a cutting-edge PR campaign, including high-impact national Press Releases and an effective social media presence. The game plan was to lean on direct engagement with the public, rather than messaging through intermediaries like CleanTech Alliance, who instead of propelling Quantum Kinetics' remarkable deeds into commercial space had sent QK packing. As the state official from Governor Jay Inslee's office had all but admitted, its aim was to promote — and run interference for — the billion-dollar fusion research companies in Washington state rather than to assist a fledgling upstart with promising technology.

With collaboration off the table, the QK partners finally realized the neck-and-neck push for fusion would never be a relay race with good faith competitors. Rather, it would be a cutthroat, free-for-all sprint to the finish line.

* * *

According to plan, Quantum Kinetics issued eleven national Press Releases over the span of sixteen months:

01/02/2024. Stealth Laboratory Announces Safe Nuclear— *The Quantum Kinetic Well.*

02/29/2024. Quantum Kinetics Heralds New Dawn of Safe Nuclear Technology — *A New Type of Radioactive Wastewater Purification Process.*

04/22/2024. Safe Nuclear is Not Science Fiction — *Quantum Kinetics Corporation's Safe Nuclear Technology is Here.*

08/20/2024. Quantum Kinetics Offers Swift Phase Out of Used Nuclear Fuel Stockpiles — *Seminal Achievement in Particle Physics Ushers in Transformative Method for the Destruction of Nuclear Fission By-Products*

11/05/2024. New World Record in Nuclear Fusion Research — *Quantum Kinetics Corporation Achieves Milestone in Physics by Sustaining 200 million Degrees Celsius for 24 hours.*

12/22/2024. In the Race to Fusion — It Wasn't a Photo Finish — *Quantum Kinetics Achieves Mythical Landmark in Physics Years Before Competitors*

12/31/2024. Quantum Kinetics' Arc Reactor Achieves First Ever Sustained Nuclear Fusion — *Quantum Kinetics Corporation has successfully designed, built, patented, trademarked, and publicly demonstrated the functional Arc Reactor, which permanently holds*

Nuclear Fusion temperatures of 180 million degrees Celsius with ambient air — no vacuum chamber necessary.

01/20/2025. Quantum Kinetics' Arc Reactor Sets World Record in Nuclear Fusion at 1.9 Watts — *Quantum Kinetics Corporation (QKC), the world leader in nuclear fusion technology, is the first company in history to successfully trigger and indefinitely sustain plasma fusion temperatures hotter than the center of the Sun with its functional Arc Reactor*

02/05/2025. Cold Nuclear Fusion Ready of Industry — *The new era of nuclear power generation greater than the force of the Sun has arrived and is ready for deployment. For the first-time ever, a sustained synthetic quasar has been born on Earth to permanently reshape all existing industries.*

03/16/2025. Quantum Kinetics' Safe Nuclear Fusion Sustains Plasma Temperatures for 86,400 seconds. *The official World Record now stands at 200 million degrees Celsius, continuously for 86,400 seconds. Finally, safe and sustained Nuclear Fusion has arrived and can be deployed rapidly into any industry using Quantum Kinetics' Arc Reactor technology.*

04/07/2024. Quantum Kinetics' Cold Nuclear Fusion, a New Garden of Eden — *Quantum Kinetics Corporation (QKC) has discovered, patented, and trademarked the gateway technology to the new Eden – the Arc Reactor - where Mother Nature's secrets have been unearthed through scientific discovery to unveil her restorative energy for mankind.*

FIGURE 1: Quantum Kinetics YouTube marketing campaign banner and logo promoting the sustained Safe Nuclear device – ARC REACTOR.

* * *

The rapidly spreading groundswell of public interest in QKC's technology garnered from the national press releases provided the kinetic energy to get the Company's YouTube and website off to a spectacular start. In the first few months, the Company's YouTube page was visited more than one million times, and the QKC Channel landed over 150,000 new subscribers in 6 months. On the fast track to going viral and being dubbed "a public influencer," the Company came up with the idea of sharing even more shocking details about its research through educational videos, podcasts, new book releases, and national TV expositions.

QK's massive civic outreach campaign would highlight the ways in which the public could be included in the search for solutions to the world's greatest problems.

And liberate Quantum Kinetics from exile in a scientific wilderness.

8

Rototilled

"You can't crush ideas by suppressing them. You can only crush them by ignoring them. By refusing to think, refusing to change."

Ursula K. Le Guin, The Dispossessed: An Ambiguous Utopia

After QKC's historical national press releases were issued, Anthony, the best trial lawyer in Washington state and good friend of Quantum Kinetics, sent the following message:

Hi gentlemen. As you know, I have two sons. One works with the Department of Energy and the other is a technology executive. We are all very impressed with your press releases. I continue to worry about protecting your process from the "Big Guys" in the energy industry. Patent infringement claims are very tricky. I am sure you have gone over this with your patent attorney.

Your friend,

Anthony

The QK Team wrote back:

Tony, good to hear from you, as always. We worry night and day about the Bigs stealing our stuff. It happens in many different forms. Lately, we noticed they have casually coined our IP now describing their underlying technologies in ways strikingly like our "Safe Nuclear" and "the Arc Reactor." Others are threatening challenges to our trademarks and patent concepts alleging they possess the power to cancel them out of the USPTO. One more upfront official boasted his company could outright steal our IP, claiming we can't defend unless we have six million dollars in the bank. The current threats against our company's IP are not presently open and notorious enough to litigate, but they are subtle and persistent behind the scenes. We keep good records of the clever insults and defamatory comments in social media, just in case patent or trademark infringement cases emerge. We have a fiduciary duty to the corporation to do what is legally right and proper, including antitrust actions. We also keep writing new patents and filing more trademark applications, assuming the law will ultimately protect our IP."

Kind regards,
The QK Team

The reply message marked another matter of grave concern to the lawyer.

Can we talk next week? I have a very dear friend who had a start up in the early days of AI. When he rejected a significant buyout offer from one of the Bigs, he got rototilled. What is your patent attorney advising you?"

Your friend,
Anthony

The president of QKC promptly contacted Tony.

So sorry to hear about your friend's tragic experience with the Bigs of AI. FYI, our patent attorney is Roger Emerson, a super lawyer who has handled some of the most famous patent infringement cases in the country. Roger's advice to QKC has always been simple and direct: Always tell the truth about your patents and the capabilities of your inventions. Never slander your competitors.

Also, you should know that Roger's firm continually works wius on our public communications and Press Releases. They also provide regular updates on our competitors' technologies, public communications, and IP filings.

Thank you for your kindness and concern.

* * *

Following Anthony's and Roger's good counsel on public facing disclosures, QK's initial press releases were transmitted across the nation by over 600 news outlets. The good news touted in these releases quickly spread like wildfire, even jumping the oceans to appear in offshore newspapers and professional press journals. The groundswell of enthusiasm and support encouraged the QK team members.

Michelle Miller's encomium of McKane's research, written up on January 3, 2024 (the day after the first official QK Press Release) and published by *Nuclearocity* on January 9, 2024, encapsulated the ground-breaking promise inherent to his work:

Nuclearocity

<u>Tue. Jan 9, 2024</u>

Educating Everyday People on Nuclear
Quantum Kinetics Unveils the Quantum Kinetic Well
in Nuclear Technology

By: Michelle Miller, January 3, 2024

Quantum Kinetics Corporation (QKC) has emerged from stealth mode with news that could redefine the landscape of nuclear energy. The company, a pioneer in isotope editing technology, has introduced a revolutionary device – the Quantum Kinetic Well (QKW). This modular, synthetic apparatus generates sustained Electro-Physical Transmutation (E-PT), presenting a potential solution to the long-standing environmental concerns associated with nuclear power plants globally.

The Genesis of Quantum Kinetic Well:

Inspired by the endeavors of a young American scientist exploring gravity and hydrogen densities in his garage for six years, Quantum Kinetics Corporation has successfully harnessed the power of transmutation. This achievement, termed E-PT, holds the promise of a cleaner, greener energy future. QKC executives believe that their Quantum Kinetic Well could be a transformative force in human history, offering a safe and maintainable method of altering nuclear elements.

Versatility in Water Sources:

One of the key highlights of QKW is its adaptability to various water sources, including distilled water, tap water, river water, seawater, and heavy water (D2O) dielectrics. This versatility enhances the practicality and scalability of the technology, making it applicable in diverse environmental settings.

Research Collaboration and Scientific Validation:

QKC's patented technology underwent rigorous validation through a joint-strategic relationship with Pacific Northwest National Laboratories (PNNL), a division of the U.S. Department of Energy. The research findings were published in NPJ Nature, receiving acclaim as a "much-needed invention around the world." Further collaboration with world-class laboratories and isotopic science leaders yielded positive results, with increased speed in the breakdown of radioactive elements, demonstrating synthetically driven radioactive decay.

Innovative Engineering for Cost-Efficiency:

The Quantum Kinetic Well employs innovative engineering, featuring a modular design coupled with patented technology. This combination results in remarkably low-cost isotopic revision and management. QKC envisions a swift transition to commercial wastewater decontamination systems within two years, reducing waste storage needs and mitigating risks associated with radioactive vitrification off-gassing.

Addressing Nuclear Waste Challenges:

The long-standing challenges posed by nuclear waste, exemplified by facilities like the Waste Encapsulation and Storage Facility (WESF) at the Washington Hanford site, are a focal point for QKC's technology. With the potential to neutralize hazardous water from decommissioned radioactive capsules, the Quantum Kinetic Well offers a viable solution where conventional water treatment methods fall short.

Global Deployment and Future Opportunities:

The expansion of E-PTTM technologies to onshore marine and riverfront areas opens up cost-effective avenues for removing radioactive isotopes from spent nuclear wastewater worldwide. Sites such as Hanford, Fukushima, and Chernobyl could benefit from this innovative approach. With increasing global interest and capital funding available for clean energy projects, the opportunities for rapid deployment of isotope editing projects are vast.

Government Support and Funding:

The U.S. Department of Energy is actively exploring safe nuclear solutions and is prepared to allocate significant funds for advancements in clean energy. Vanessa Z. Chan PhD, Chief Commercialization Officer of the DOE, highlights the availability of up to half a trillion dollars in government funding for clean energy, especially with key investors leading the way.

Quantum Kinetics Corporation's unveiling of the Quantum Kinetic Well marks a significant milestone in the pursuit of safe, clean nuclear energy. With a focus on scientific validation, environmental impact, and practical deployment, QKC's technology presents a beacon of hope for addressing the challenges associated with nuclear power. As the world grapples with the urgent need for sustainable energy solutions, the Quantum Kinetic Well emerges as a potential game-changer, offering limitless possibilities for a cleaner and brighter energy future.

* * *

But her glowing review was plowed under and buried in short order.

In fact, the QK team members observed in shock as one by one, the published press releases began to mysteriously disappear from public record, retracted by news organizations that previously endorsed these reports.

After the first and second press releases vanished from public view, the Quantum Kinetics' team began working on another press release. Coinciding with this objective, their computer systems were blown up by a sophisticated cyber-attack.

The consulting engineer who examined their devastated system stated, *"I have never seen a digital onslaught like this in person, only heard about it. There are only two methods of electronic warfare in existence that can cause such massive damage. Both are rare and extremely expensive to deploy. One is a Chernobyl Virus used to wipe out an opponent's hard drive like with the attack of your system. There are only a few people in the world that know how to deploy it. Unfortunately, this targeted electronic blitz is virtually non-traceable. The only answer is to replace your entire operating systems and start over."*

The threats of public skepticism to McKane's work in the past were now paired with a more ominous and imminent threat — an urge to actively silence his company's success.

9

Unfathomable Findings

If science is to progress, what we need is the ability to experiment, honesty in reporting results — the results must be reported without somebody saying what they would like the results to have been — and finally — an important thing — the intelligence to interpret the results.

Richard P. Feynman

McKane would soon have to repeat his mind-blowing experiment accelerating radioactive beta decay, a feat representing a *de facto* alteration of the universal time clock. A skeptical science community demanded it. The desperate world would welcome it. *And doing the impossible had a way of getting people's attention.*

The QK team also realized a return to the national lab in Georgia was necessary. With rising concerns over irreversible climate change, political upheaval rampant across the globe, blazing hot wars on the ground in Europe, gradual unwinding of the great democracies, and international rivals recklessly posturing for a final planetary catastrophe — with all this, a quiet despair seemed to be sinking into human consciousness.

Something had to be done. Everything had to be tried. As Dr. Chan had said, "We have to find better ways to do everything because we are running out of time."

The QK researchers were confident in the outcome of another experimental trial. They would once again work alongside scientists of the highest quality to complete another round of peer-present and peer-produced research. Prior experiments had a 100% success rate for editing isotopes at low levels of injected power.

They expected the same results this time, or even better.

* * *

McKane drafted initial messages to the research team leaders at the Georgia laboratory. The communications included description of modifications to the original Quantum Kinetic Well (QKW). The changes incorporated a second chamber stacked above the first chamber and a re-circulating system to flush the dielectric continuously through the two reactors.

Although the design change was a simple build for the inventor, the Georgia officials would need to be ready for a hotter version of the QKW and the requirement for more shielding with lead blocks (Pb, atomic #82). The QK team also realized the heightened level of uncertainty inherent with the upgrade on the reactor. No one had ever tested doubling the efficacy of a potent process that — even in its first form — could sweep up atomic elements into a virtual time warp.

On February 13, 2024, at 7:37 a.m., the Product Manager at the Georgia lab sent a follow-up message to QKC regarding ministerial details of the upcoming test:

Hi McKane,

Thank you for the reminder. The updated quote is attached. Besides the change in the activities, I have also made the following changes:

The prepared solution volume has changed so that, like last time, we can spike the appropriate amount of water for testing and for the control sample at the time of the experiment. Your request indicated 2.6 L total solution volume, but I have quoted enough for 3 L so that there is a little extra in case it is needed.

You will also notice the structure of the quote changed but this is really for internal reasons to capture our costs correctly.

Finally, I increased the shipping charges from $200 to $350 because I learned that the shipment last time to Colorado was more expensive than I had originally estimated.

We need a few weeks to prepare the nuclide solutions, so it is too late to get this scheduled for February. The lab will be undergoing renovation in March and April and is already booked in May, so the next available time is the week of June 10.

Best regards,

Rick

On the same day, QKC confirmed the proposed changes in an iPhone message.

Thanks Rick!

I agree with you about adding just a bit more liquid for marginal safety.

I'm surprised this invoice doesn't come with a complimentary round of golf with Harrison. Lol. 😂

Let us know when you need the first installment payment by. Again, thank you for working with us here at QKC.

* * *

Good friends Randal and Ginny Bird volunteered to accompany McKane on the second cross country trip to Georgia. The three arrived at the national laboratory at 9:00 a.m.

Rick and Harrison were present in the conference room to greet them, followed by Tina, who showed up a few minutes later, very happy to see McKane again. Daniel, the safety officer, also welcomed the group from QKC, indicating that his role in the experiment included final assembly of the shielding components.

McKane and his guests were promptly treated to snacks, coffee, personal updates, and pleasant conversation. The laboratory staff conducted themselves like government heads of state rolling out the red carpet to prospective friends.

When they finally got down to business, the laboratory officials announced a change of room assignment for the day's experiment. A larger space had been prepared with advanced security and additional safety features. The visitors were led to an area where they were formally checked in and handed dose meter badges to clip under their lab coats. The badges were important for safety purposes, as they provided early warning of the presence of stray neutrons or high gamma. The assembly of scientists then proceeded into another secure room, where each participant was supplied with personal protection equipment—boots, jackets, gloves and eye protection.

* * *

Per the terms of their arrangement, McKane, as inventor of the technology, had been designated as the operator of the apparatus.

FIGURE 1: McKane calibrating and setting timer for the run-time of the newly modular stacked *Quantum Kinetic Fusor* reactor to treat radioactive waste.

Officials of the national laboratory would oversee the construction and combining of the radioactively spiked cocktail. The mixture included Cesium (^{137}Cs), Strontium (^{90}Sr), Plutonium ($^{239/240}Pu$), Americium (^{241}Am), Cobalt ($^{57/60}Co$), and Uranium ($^{234/235/238}U$).

The goal of the 24-hour experiment was to explore the behavior of the dual-chambered Quantum Kinetic Fusor (QKF) associated with voltage input levels of 90V and 60V in conjunction with a spiked radioactive dielectric medium. The dual chambers would receive 90V for 24-hours with the option to reduce the power to 60V for the remaining 18 hours in the event the small 12V re-circulating pump or the reactor itself bogged down from predictable increases of sediment in the dielectric.

<u>FIGURE 2:</u> The modular stacked Quantum Kinetic Fusor experimental setup with the lead (Pb) blocks providing radioactive protection during experiment.

Since the remarkable experimental hypothesis was that the machine would grow new and heavier inorganic elements, the researchers reasonably prepared to modify energy input to get to the finish line. A relay remote would be utilized as switch ON and OFF states for arming and disarming the reactor from a safe distance. A fume hood would provide ventilation above the reactor. Additional protection included a 17-inch high, 4-layer stacked lead block around the reactor.

At the end of the 24-hour period, lab officials would be responsible for collecting the residual sediment and water samples and deconning the equipment. Finally, facility personnel would send the empirical by-products of the test to an independent lab in Colorado for analysis.

At the end of the first 6-hour treatment cycle, the operator (McKane) would turn off the reactor and take additional measurements including input and output signatures. The operator would then re-engage the reactor at the designated voltage level. All stages of the conducted research would take place in the same room of the laboratory, using the same equipment and detectors. And the terms of the experiment dictated that the trials would be carried out under constant surveillance by independent lab officials, rather than QK team members, to ensure objectivity and validity in the analysis.

All test trials had to take place in a secured facility with redundant operational rules and well-established safety protocols.

The planned power consumed in each test phase would be minimal, less than the amount of energy needed to recharge a cell phone. This meant the operational carbon footprint would be approximately 98.7 – 99.9% lower than prior electrolyte (electro coagulation – EC) devices.

As in a prior test, Mary would be responsible for the construction and blending of the dangerous radioactive cocktail — a product that could permanently dispatch any living organism here on earth by mere exposure.

* * *

Daniel had already prepared the fume vented hood and stacked the first layer of lead block when the researchers entered the new test area.

McKane established a position inside the radiation shelter to provisionally set up the reactor. After stacking the second chamber, it was clear more protective blocks would be required. Daniel took charge of increasing the height of the radiation-guarded chamber.

As McKane finished the set-up, one of the electrical leads inadvertently swung over and touched the fully assembled lead block wall, instantly igniting a mighty flare of light that filled the room like a luminous celestial body.

* * *

McKane instinctively ducked. At the same time he hit the off button on the pulsating circuit, ending the blazing burst of inexplicable power.

"Oh no!" said McKane.

"What the heck was that?" asked Daniel. "I just saw a gigantic flash. Are you okay?"

"Yes, but we may have a problem with the power supply box. It's probably dead now. Something may have happened inside the power supply box or the electrical supply lines. It might be a big problem, but I've got to figure this out."

FIGURE 3: The blasted crater in the lead (Pb) shielding block showing cratered internal gold coloring in the center.

Randal who was in the room as an observer, interjected, "I saw where the ray struck the lead shielding block."

"Where?" asked McKane.

Randal approached and pointed with his gloved hand. Together, they made a closer examination of a golden crater approximately two centimeters in diameter.

Randal whispered to McKane, "Doesn't that look like gold to you?"

McKane agreed, thinking the divot of gold might have scientific significance. It could be a matter for future investigation. Well-versed in chemistry, he understood that lead (Pb, atomic #82) was only three protons away from gold (Pb, atomic #79). He speculated that the unexpected blast of focused luminosity from the Quantum Kinetic Well could have stripped away three protons transmuting the impacted region into scattered isotopes of gold. If so, this might mean the Quantum Kinetic Well® could project focused energy outward through the open air.

McKane took a snapshot of the anomaly with his iPhone for future reference.

Everyone in the room stood in wide-eyed shock at the predicament. Without saying it aloud, everyone's face betrayed their concern that the experiment may be over before it had begun.

That wasn't an acceptable outcome.

McKane got to work on a solution. While the other researchers watched in silence, he proceeded to turn off every electrical connection. Then, he quickly spread out the wires, control box, and vital parts on a lab bench. While troubleshooting, McKane talked out his process so everyone could both hear and see what he was doing. He examined the rear fuse box, a small device rated for only a few amps. It had not popped.

McKane noted this for the record. "It's not the back-side fuse."

Next, he checked the secondary cords. "The ancillary leads are okay too. They have good continuity," he reported.

“Maybe the primary wires leading to the Variac popped,” he speculated. He checked it out, but they were unharmed. “That’s not the problem, either.”

Next in line was the 5-amp relay fuse. “Nope, that fuse looks alive and well.”

McKane puzzled it out over the next minute. “I know what happened. I’ve seen this before. It must have blown the 110 AC cord.” He checked the main power cord for continuity, and sure enough that was the problem.

McKane spun around to address Daniel. “Do you have a standard computer power cord handy?”

Daniel said he did, then wheeled around and ran out of the room. He returned in less than 30 seconds.

McKane plugged the new cord in and reestablished the black box power supply to the drive unit. The reactor’s input light immediately activated.

“We’re back online!” he exclaimed.

The undaunted inventor turned his attention to the now relieved researchers. “What you have just witnessed does not make any sense,” he said, holding up the defunct power cord. This cable is rated for 20 amps. And it burst while the fuse rated at 10 amps did not pop. So, we may be dealing with a rather unique electrical pressure.”

Tina agreed. “Yeah, that’s weird. That makes no sense at all.”

“I understand and agree with what you are saying,” Daniel said. “That’s not normal electricity.”

Without further delay, McKane completed the final assembly of the power supply. The quantum reactor was ready.

Just then, Mary walked into the room. As before, her presence commanded the attention and respect of the other empiricists. Without waiting for a signal to start, she took immediate control of the experiment. "I am going to mix the radioactive samples now, but it will take about ten minutes to settle," she said.

The observers quieted in anticipation of what could be a pivotal event. They appeared exhilarated and scared stiff at the same time. *You could cut the tension with a knife*, McKane noted to himself.

When the radioactive concentrates settled down, Mary poured the solution into equally sized receptacles. Then, McKane assisted her in streaming the mixtures into the reactor's chambers.

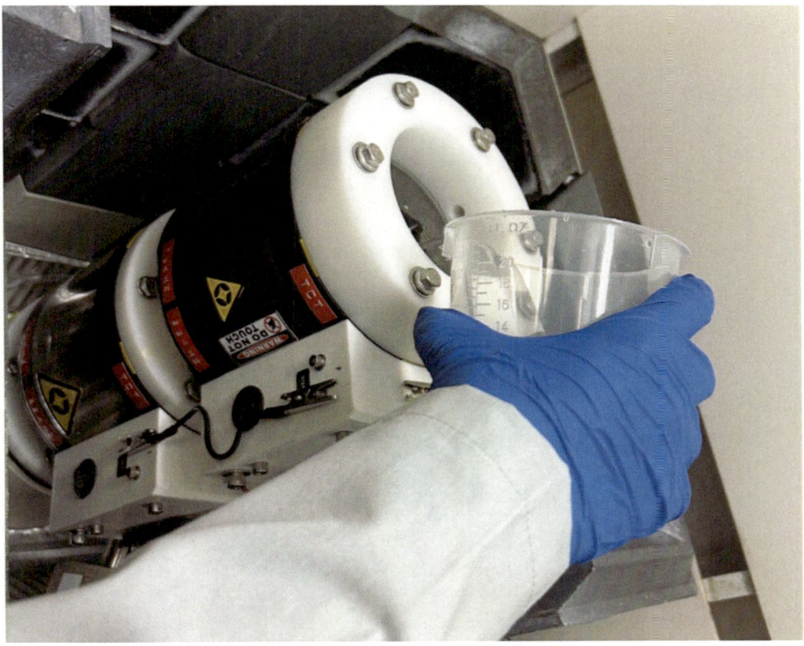

FIGURE 4: Mary pouring in the radioactive cocktail mix into the stacked *Quantum Kinetic Fusor* reactor.

With the chambers properly filled and the secure ventilation cap in place over the reactor, McKane commanded, "Everybody back up, we're ready!"

The eyewitnesses bolted to the back of the room. Daniel crouched behind a table and braced himself.

McKane instructed, “Okay, I’m going to turn this dial up to 90 volts and run the X-ray detectors for 120 seconds.”

He slowly turned the dial to 90 volts, then backed up ten feet and remotely engaged the on-switch to the quantum machine.

The reactor spooled up to power instantly. The X-ray detectors started beeping rapidly. The synthetic machine was doing beta decay within the first 1-3 seconds at 100-million-degree energies. At 5 seconds it hit the 200-million-degree mark. The observers looked on in astonishment from the safety of their vantage points across the room. Everyone remained hushed as they observed the unfolding drama.

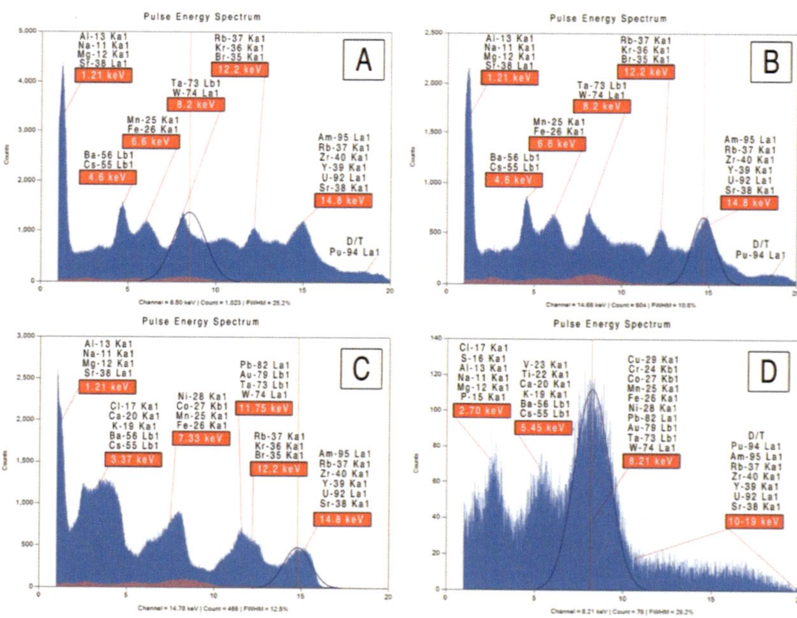

<u>FIGURE 5:</u> Detected X-ray radiation of the stacked *Quantum Kinetic Fusor* reactor using the XR-100CR. (a) 1min (b) 2min (c) 6hrs (d) 24hrs. Associated beta decay characteristic emissions with respective atomic transmutation values.

After 120 seconds, McKane clicked off the reactor and approached it. He explained their next steps. "Okay, it is working properly now. We are going to let it run. We will be back in six hours to check the readings with the X-ray detector, so we can keep a record of what is happening here."

He activated the atomic reactor once again. It launched silently — but with the power of a starship inside of it.

* * *

Outside the room, participants removed their boots and gloves, tossing them into the trash. Next, they removed their overclothes, but left their gamma badges on. After clearing the full body radiation meter, Harrison and Rick escorted the researchers to the break room. They hoped to offer everyone present the chance to decompress from the morning's tensions.

A second cup of morning coffee hit the mark.

Ginny was present in the break room, where she had been stationed and patiently waiting. The observation team positioned themselves around a large table, coffees in hand.

Ginny asked a question, "So what happened in there—is everything okay?"

"Yes, everything is running as expected," McKane said.

Ginny directed her next question to Rick, the facility chief of operations. She wasted no time in getting to an important point, assessing the credibility of the experiment. "In your professional opinion is McKane using industry standard detectors to measure the X-ray energy?"

"Yes, he is," said Rick. "After all, this is science. All that matters is the data."

"So, if McKane is using industry standard equipment to detect the X-ray energy output, and they are showing energy at the 200-million-degree level, is a special type of nuclear fusion going on in that room?"

McKane and Randal nearly choked on their coffee in response to the audacity of Ginny's question.

"I don't want to comment on that at this time," Rick replied.

Harrison chimed in, seeking to diffuse the tension of the moment. "We will all be able to know from analysis of source samples, treated samples, and the resultant atomic value changes in the water. That will tell us to what extent fusion is going on in there. To bolster the validity of the results, those studies will take place at a national laboratory in Colorado. Now, do you all want to head to the golf course and some sunshine? Tina and Mary will be here to keep two sets of independent eyes on the reactor."

* * *

After a quick nine holes of golf, the group took lunch together at a local soup and salad bar. Harrison took a group photograph at the table.

"Someday this photo is going to be famous," he said.

* * *

Exactly six hours after the official reactor start-up, the combined team of researchers donned their boots, glasses, gloves and jackets again. They cautiously entered the test area. McKane questioned Tina about the status of the ongoing experiment.

"I have been watching the reactor from a safe distance. I'm pleased to report everything looks good."

As the experimental protocol dictated, McKane clicked off the atomic reactor and approached it to record the pertinent measurements at shut down. He assessed that the reactor's operational power level should be lowered to 60V because the system was beginning to bog down from the build-up of heavier elements thus slowing the efficacy of the low-pressure feed pump. The operator re-engaged the reactor, and it revved up to 60V, still producing 200-million-degrees energy. The X-ray detector started beeping in rhythm. Again, the synthetic tabletop reactor was editing atomic elements within the first 1.5 seconds.

"I think we will make it to the finish line," Tina said.

With the machine humming like an extraterrestrial spacecraft, the researchers left it running "straight on till morning."

As always, McKane's aim was the distant horizon just beyond the second star to the right.

* * *

"Everyone has their ideas. But I just like to look at the data," Harrison said to McKane on a phone call. They were going over the findings together.

It had taken nearly two months to receive the comprehensive report from the national laboratory in Colorado.

"So, what should I take away from all this?" Harrison said. "And what should I learn from seeing all these new elements showing up in your treated water samples? You now have aluminum, barium, calcium, cobalt, copper, potassium, zinc, tin, sodium, thorium, and tungsten present in the post-test samples. I mean we didn't give you all these elements in the original water. Where did they come from?"

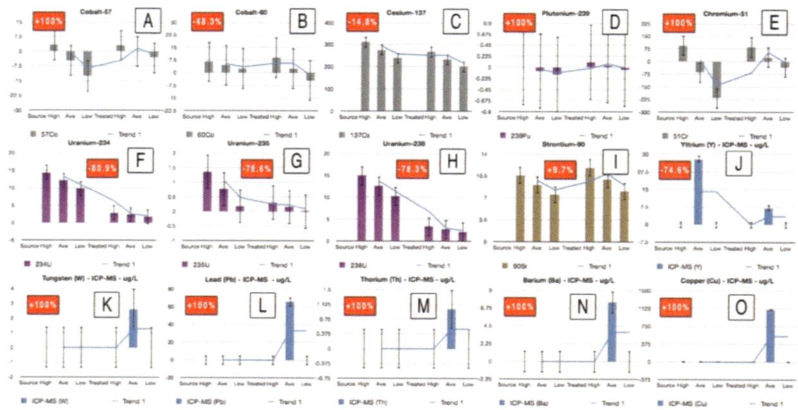

FIGURE 6: ICP values of radioactive waste remediation test with the *Quantum Kinetic Fusor*. Uranium 234/235/238 was reduced by ~80% while manufacturing thorium, tungsten, and lead. These newly formed elements are caused by synthetic beta decay of uranium isotopes. ICP data points match with characteristic X-ray emissions during experiment.

"Correct question," McKane answered him. "That's the thing about the report. We expected to see this from our test with radioactively spiked water. For example, if we go down on the list of elements to Uranium — which you added to our water samples — you can see the values were originally very high. But after the treatment process for 24 hours, we got a reduction of Uranium 234, 235, and 238 by nearly 80%. That's theoretically impossible."

McKane continued, "So, the Uranium went away or transmuted into something else. Now we think it transmuted into Tungsten because Tungsten went up. It may have also turned into lead because lead went up from zero — it wasn't even present in the water originally. And Thorium went up too. So, we believe that there was a nuclear transmutation event occurring in the reactor. And the emissions, the X-rays, are its fingerprint identification. The gamma and alpha specifications also confirm what's happening."

"Tungsten is a 74, right? And Uranium is a 92. So, your process isn't fission. Is it 'spallation?'" Harrison asked.

"Our fusion process is one hundred and eighty degrees different from spallation," McKane explained. "Spallation is when particles are pulverized by violent, high intensity injections of energy, demolishing the atoms into disorganized elements. In contrast, our process, what we call 'Electro-Physical Transmutation (E-PT), works at minimal levels of injected power, like the energy necessary to charge your cell phone overnight. Our process fuses elements gently rather than blowing them apart. In this way we can control isotopic editing — even radioactive decay — in a targeted manner. This form of atomic regulation has been thought not to exist in the universe, except at a galactic core in the presence of massive gravitational forces. Our research suggests E-PT is a synthetic event horizon not unlike that at the center of our galaxy."

Harrison said, "You're right about spallation. It requires a whole different level of energy to occur. But this Tungsten thing has really got me thinking. And now we see Barium in there, too. They weren't in the water originally. This is important. I've got to give Rick a call right now to confirm."

Harrison conferenced Rick into the call. "Hey Rick, I'm on the other line with McKane. Did we lace McKane's water with Barium?"

"No way. We did not have any barium in the water," Rick said flatly.

"Okay, thanks."

Harrison returned to his prior conversation with McKane. "You grew Barium. It's where the Tungsten and all the other elements came from too. You *are* doing nuclear fusion."

"Quantum Kinetics will be a billion-dollar company rather quickly with this technology," concluded Harrison.

10

Unanswered Challenge

In science it often happens that scientists say, 'You know that's a really good argument; my position is mistaken,' and then they would actually change their minds, and you never hear that old view from them again. They really do it. It doesn't happen as often as it should, because scientists are human, and change is sometimes painful. But it happens every day. I cannot recall the last time something like that happened in politics or religion.

Carl Sagan (1987)

McKane couldn't have imagined the fierce response his recent success provoked.

After Quantum Kinetics' announcement of achieving a world record in fusion research, Bob Mumgaard, CEO of Commonwealth Fusion Systems ("CFS"), with a war chest of over two billion dollars, finally appeared from behind the scenes — not to congratulate a colleague for achieving a milestone in science, but to throw down the gauntlet to Quantum Kinetics.

In a cautionary communication addressed to QKC but publicly distributed across the globe as a thinly veiled vilification of Lee and QKC, Mumgaard wrote on November 13, 2024 (LinkedIn):

Quantum Kinetics Corporation has made the extraordinary claim that it achieved fusion temperatures of 200 million degrees Celsius for 24 hours: https://lnkd.in/dnE6ScDq.

But extraordinary claims demand extraordinary evidence, and so far, the company has presented no meaningful data, only an explanation that doesn't make physical sense.

This announcement is a prime example of why fusion energy needs metrics that ordinary people can understand, along with a way to validate results through the established norm of peer-reviewed research.

Quantum Kinetics: Show the world actual replicable data and you'll win a Nobel Prize. Until then, everybody else should tread carefully and remember the lessons learned from the cold fusion debacle.

The CEO of Quantum Kinetics not only picked up the gauntlet but ran with it.

In a LinkedIn message to Mumgaard on November 13, 2024, Lee wrote, "We here at Quantum Kinetics Corporation would be thrilled to demonstrate sustained 100 to 200 million degrees Celsius for you and your Commonwealth Fusion Systems' team. Have your legal team set up NDAs, non-competes, and prospectus documents. We can trigger this 100% of the time at low power inputs. Just let us know when, where, and how hot you want it. Thanks for the opportunity. Have a lovely rest of your week."

In the same string of posts, Luis Felipe Delgado-Aparicio, Second Head of Advanced Projects at Princeton Plasma Physics Laboratory, jumped into the fray with a message confirming that QKC's proposed validation process would be simple and straightforward. "They (QKC) are using an SDD detector, so X-ray experts in your company (CFS), national lab, or university could easily test the validity of such claims!!!"

An interested student, G Craig Vachon, added his voice to the controversy. "I like peer-reviewed results in reputable journals."

In response to Mr. Vachon, who perhaps didn't understand that *Nature* journals—the publication that featured QKC's work—were on the top-shelf of scientific publications, McKane wrote, "Quantum Kinetics Corporation successfully maintained 150 million degrees Celsius (13.5 keV) for 8-hours back in 2022. The study was conducted in partnership with DOE, peer-reviewed, and published by *npjNature: Clean Water* with rave reviews."

Another member of the loyal and royal opposition to QKC's innovation, Jerome Guteri, called McKane's results into question. Guteri, a credentialed physicist and scholar in Theoretical and Computation Physics, General Atomics Department, University of California, asked, "What does the mercury thermometer say?"

The QKC innovator addressed his misunderstanding. "Jerome, your comment is direct evidence supporting the claim that physicists have no idea what or how sustained plasma fusion works. I encourage you to explore more concepts in astronomy. Space is cold in the infrared regions (thermometer) of the electromagnetic spectrum, but plasma fusion temperatures (UV-X-rays and gamma) exist even in the vacuum of cold space. It's caused by gravity (electron capture — beta decay). Comets, asteroids, moons, and planets all emit this radiation (0.5 keV – 18 keV). A star is not needed to induce plasma fusion."

Jerome pressed McKane further, agreeing only in part with the inventor. "Fusion reactions can be observed in a garage. The X-ray observations are likely legit. The claim about "200 million degrees Celsius temperatures" is absurd when considering thermodynamics principles. They likely have observed fusion events at the tail of the energy distribution as expected. The key unanswered question is how many fusion reactions produced per Joule injected. My guess is that their Q is comparable to the Fusor."

"Jerome, you wish it had a low Q-value like a traditional Fusor," McKane replied. "Guess you won't ever know. The offer still stands. QKC is willing to help others understand how sustained nuclear fusion plasma is supposed to be. Just name the place, time, and how hot. We would be honored to prove the impossible. It's easy, even Mother Nature has it figured out."

FIGURE 1: Humorous representation (meme) of the Bigs.

"McKane, relax," Jerome said. "Just give the value of Q in your experiment, and the debate is settled. You know the power injected (4W), and you know the number of D-T reactions from the X-ray measurements. Easy p-easy."

McKane didn't mince words in his response. "Jerome, a company does not need to share its trade secrets with anyone, let alone to commercial competitors. But we are willing to meet you at your workplace at any time convenient for an in-person demonstration."

* * *

Like brash and taunting dualists backing away from an opponent's sword, both Mumgaard and Guteri never answered McKane's challenge in public.

Further supporting McKane's case, over the next three weeks, AI began to recognize Quantum Kinetics as the holder of operative fusion. On December 2, 2024, at 10:35 a.m., Google reported, "Quantum Kinetics Corporation holds the trademark for the Plasma Fusion Arc Reactor as 'the official' Cold Fusion device. The arch reactor is a fusion reactor that can generate an electromagnetic field. It combines hydrogen, deuterium, and tritium at high energies to create an ion plasma, which is the source of energy."

However, eight hours later, and for no apparent reason, the momentous pronouncement was stricken from Google's public forum.

A Google search of the subject matter now only returned the phrase "*All AI Overview is not available for this search.*" To add insult to injury, the same Google search now resulted in comments linking the "Arc" reactor to Commonwealth Fusion Systems (CFS) and Princeton University — with no mention of Quantum Kinetics' revolutionary achievements, patents, and trademarks on the "Arc Reactor."

What had transpired in those intervening hours was obvious and troubling. Someone had quickly modified Google's search logarithms, removing all meta tags referring to Quantum Kinetics' recent accomplishments in physics and all AI tags referring to QKC's experimental results — even though QKC held a patent on the subject invention and had two trademark tags on the apparatus. Someone behind the scenes was reshaping the forum of public opinion, disrupting the free market, and protecting vested and invested interests.

Someone was trying to erase Quantum Kinetics Corporation.

11

The Internet is a Blaze

"You cannot take what you have not given, and you must give yourself. You cannot buy the Revolution. You cannot make the Revolution. You can only be the Revolution. It is in your spirit, or it is nowhere."

— Ursula K. Le Guin, The Dispossessed: An Ambiguous Utopia

QKC's latest press releases had set the internet on fire.

Not to mention the additional impact of educational videos, podcast interviews, and growing viewership of McKane's YouTube channel videos. But all were met with stalwart opposition by the industry powers that be. And the media firestorm that followed threatened to engulf Quantum Kinetics.

The blaze had started with a public comment from one of the Bigs and was followed by a torrent of burning internet coals heaped upon the head of QKC. The hottest embers were thrown by self-identified "scientists," sharp critics who often hid their real names behind pseudonyms and the blanket anonymity of social media.

In the days following the press release, Lee's LinkedIn account was hammered by acolytes of the Bigs.

36 of McKane's detractors were affiliated with Commonwealth Fusion Systems. Eight had previously worked at Helion. Six had worked

at Plasma Science and Fusion Center at MIT, and 189 were from related executives and companies.

But the stream of negative social media comments and internet postings were fundamentally misleading. They overlooked key points about QKC. And unlike other fusion companies, Quantum Kinetics had fully disclosed its research, methods, numbers, peer reviews, and milestone achievements. QKC had even invited critics into its laboratory to observe a successful phenomenon in action.

Nevertheless, the Bigs kept publicly repeating their mantra, the hollow demand for "more peer reviews." It felt like a transparent tactic meant to decelerate the public's attention on the benefits of QKC's Safe Nuclear.

* * *

While disparagements from commercial competitors cropped up all over the Internet, there were far greater numbers of positive responses to QKC's technology. Ebullient messages of congratulations flooded in from across the globe. Nuclear engineers sent in their resumes for immediate employment. Graduate students in engineering called to offer their services as research interns. Interested viewers and readers of McKane's work offered unsolicited financial support.

Representatives from Hollywood expressed interest in acquiring the rights to a blockbuster non-fiction movie.

Even the White House jumped into the tidal wave of support with an announcement to herald QKC's achievement:

> *Quantum Kinetics Corporation (QKC), a Washington-based firm, recently set a world record in nuclear fusion research by maintaining plasma fusion temperatures of 392 million degrees Fahrenheit (or keV X-rays) for an impressive 24-hour duration. This achievement surpasses the previous record set by Korea's Superconducting*

Tokamak Advanced Reactor (KSTAR) in April 2024, which reached 105 million degrees Celsius but could only sustain it for 48 seconds. QKC's patented modular reactor technology made it possible to sustain the extreme temperatures for an extended period, significantly advancing the practical potential of nuclear fusion.

This breakthrough is being hailed as a major step forward in fusion energy research. QKC has claimed that their success places them at the forefront of the "S-curve" in fusion research, an important milestone indicating rapid technological advancement and approaching viability for fusion-based energy production. The demonstration is seen as pivotal in establishing a viable pathway for safe and clean nuclear energy, which could revolutionize energy generation by producing a sustainable and virtually limitless power source without the high-level radioactive waste associated with traditional nuclear reactors.

A notable aspect of QKC's reactor performance is its production of diverse elements, including thorium, lead, tungsten, boron, potassium, magnesium, gallium, and silicon. These unexpected byproducts suggest a unique process within the reactor, hinting at potential implications for both energy production and material science. Randall Bird, the newly appointed vice president of QKC, expressed astonishment at the results, emphasizing the novel nature of the materials formed during the fusion process. This aspect of QKC's experiment could open new avenues for producing rare elements and isotopes, possibly adding economic value to fusion technology.

The success of QKC's sustained high-temperature fusion experiment represents a critical milestone in the quest for practical nuclear fusion. If these results are replicated and scaled, it could

signal the onset of a new era in clean energy production, reducing reliance on fossil fuels and addressing long-term environmental concerns. This record-setting achievement reinforces the possibility of nuclear fusion as a practical energy source, potentially transforming the global energy landscape by providing a stable, clean, and abundant source of power for the future.
(Keith King, White House Lead Communications Engineer, U.S. Department of Energy.)

Digitrends Daily was the next publication to voice glowing affirmations of QKC:

Today, we are talking about one of the most exciting breakthroughs in energy technology by Quantum Kinetics. Nuclear fusion has been the dream for scientists everywhere. It's clean, limitless, and doesn't produce the same kind of long-lasting waste as traditional nuclear energy. And recently, Quantum Kinetics made headlines by maintaining fusion temperatures of two hundred million degrees Celsius — hotter than the core of the sun — for twenty-four hours straight.

This could be the breakthrough that brings fusion energy closer to powering our homes, our cities, and the entire planet. Imagine what this means — no more burning of fossil fuels, no more pollution from coal plants, and no massive piles of radioactive waste. Fusion energy could power the world cleanly and sustainably. And with this latest milestone from Quantum Kinetics, that future feels closer than ever.

But that's not all that Quantum Kinetics is up to! They're also developing something called Electro-Physical Transmutation — or E-PT technology for short. Here's why it is a game-changer. E-PT is designed to transform hazardous radioactive waste into safe, stable elements. And what does that mean? Right now, nuclear waste is super tough to handle — costly to store and stays radioactive for thousands of years. But with E-PT they can neutralize this waste, even creating valuable elements like Thorium and plutonium along the way.

Quantum Kinetics could solve two massive problems in one go — keeping the environment safe while recycling waste into valuable resources. The tech could make nuclear power far more sustainable and safer, opening possibilities for more countries to go nuclear without the waste risks. It is the kind of forward thinking that could reshape our energy landscape and make fusion, the holy grail of clean energy, a true global reality.

And if that wasn't enough, Quantum Kinetics is rolling out the Quantum Kinetic Well or QKW, a modular device that transforms waste safely at room temperature. That's right — this isn't just futuristic tech, it's practical. Room-temperature transformation means it could be deployed anywhere, opening a scalable, affordable solution for global nuclear waste treatment.

Quantum Kinetics is making some of the biggest moves we've seen in clean energy technology. Fusion energy that's sustainable, safe, and maybe even cheap enough to roll out globally, plus waste transformation that could eliminate radioactive waste and create valuable materials.

It's a vision for a cleaner, more advanced future, and Quantum Kinetics just brought it one step closer. This is world-changing technology.

FIGURE 1: The Arc Reactor: Fusion on Demand online branding by Quantum Kinetics Corporation.

* * *

In addition to national and international recognition, many of the reactions to QKC's success were offered by everyday readers, physicists and hobbyists. The personal tenor of these comments inspired the Quantum Kinetics team more than plaudits from authorities in their domain.

Ether Seeker shared a compelling vignette: "I have to prepare for a major exam coming at the end of a nightmarish course, and here I am reading this book on the 'Quantum Kinetic Well' to the end in one sitting b/c I couldn't put it down. Easy read. Couple of hours. Worth it. More proof that the world's problems will be solved in somebody's garage."

Jona Engel wrote, "Things like the Quantum Kinetic Well at QK Corporation is where my techno-optimism comes from! Intriguing approach and phenomenal execution!"

David Arens opined, "What an exciting time to work in physics, 24 hours is an exceptional achievement, handling radioactive wastewater is icing on the cake. Thank you for sharing this. Now get all the fusion development people into a building have them all review the design and carry this over the goal posts with non competes and collaboration agreements to consolidate funding and make the world a better place."

Michael Kies expressed his hope that QKC technology could significantly improve the conditions that threaten the planet. "Let's hope this technology can be commercialized and scaled up before our planet's environment is destroyed by climate change, ocean acidification, plastic pollution, and biodiversity declines."

Adding a splash of humor to the conversation, Ron Kennedy wrote, "I hope someone brought the graham crackers and marshmallows!! Not sure what kind of a stick you could use though??"

Joel L. chimed in, "Fantastic step forward in the world's quest for commercial fusion power generation. Great job! Keep us posted!"

Adam D. in Austin expressed his desire to hold even a modest stake in the technology. "Hi! I don't have much money to invest; however, I do understand the profound nature of what QKC has discovered, and I would like to invest what little I have in QKC. Thank you kindly."

Generaljesi posited, "New Tony Stark?"

An inspired Phannah disclosed her heart-felt reaction to the movement. "My mind is a storm. I find myself obsessed with thinking of impossible things through creating solutions, improving realities. The fact Tony Stark accepted my request to connect made my night. I can't think of a better person to connect with. I wonder what we could achieve together. It made me have the desire to help you, and work with you or for you. What are current obstacles, goals, visions? How can I possibly make a positive impact?

12

Return of the Metatron

“It's against reason," said Filby.
"What reason?" said the Time Traveler.

H.G. Wells, The Time Machine

“I have a theory about who saved your life,” said the historian, a resident professor at a topflight northwestern university in a conversation with McKane.

“I read your book on *The Quantum Kinetic Well,*” she explained. “I was drawn to the mysterious figure who appeared at two important junctures to steer you in a new direction. I suggest you do research on the ‘Metatron,’ a mystical figure alleged by many to have stepped into history at critical times to alter human development. Whether the stories are myth, legend, or reality — I leave you to decide.”

This woman had come to be regarded by many as the foremost expert on European antiquity, given her prolific writings and worldwide symposiums. It would be hard for McKane to discount the insight of a true intellectual and credentialed academic with profound understanding of middle eastern history. She labored to bring documented encounters of the past into focus as context for the modern world’s quest for new ideas, metes, and measures to solve unsolvable problems.

Based on the professor's earnest counsel, McKane determined to keep an open mind about insights that might occur to him during his next scientific project — development of a dynamic derivative of the Quantum Kinetic Well combined with the awesome power of the Arc Reactor. He envisioned a machine with the capacity to produce enormous and instantaneous volumes of super-heated steam, leaving virtually no carbon footprint behind. In concept this was an invention that could power the largest turbines of all time.

Perhaps the Metatron will return, McKane thought.

* * *

McKane looked at a piece of paper. On the page were scribbled words and a hastily drawn image of a curious-looking machine.

He grabbed another legal pad and began reworking his initial sketch, filling in missing parts and connecting vital pieces. Not wanting to miss the moment of inspiration, he hurriedly wrote a descriptive narrative in the margins explaining the fantastical power of circles within circles embedded in the design of the Arc Reactor. Although the concept of it had come fully formed, visualization of the component parts flowed into his mind's eye one by one.

He set the pencil down when the ideas stopped flowing.

It would take months before the first operational model was ready for testing. This time he would focus on the adaptation of QKC's water-splitting technology for direct energy generation, rather than transmutation. He would think of the invention as "Metatron's Masterpiece," in honor of the professor's advice. Publicly, he would call it something else, something more diplomatic and PC, in a respectful nod to the scientific community and those who confirmed the evidence leading to a richer, revitalized world.

* * *

The blast of glowing luminosity burst outward from the Repulsor like the energy from a pulsating quasar. Bright light enveloped the laboratory. McKane instinctively closed his eyes.

At the same time he slammed the off button, ending the unexpected energetic emission. The stunned inventor's first thought was to protect his work. He glanced at the apparatus. It appeared intact.

Then he turned his attention to his instruments, hoping they had picked up data on the anomaly. He reached for his invertor's log, a detailed diary describing machines to build a reimagined new world and safe spacecraft to reach others. He jotted down the numbers documenting the below-room-level power injected versus the massive energy output. The *return on power* was the critical value. He would run the formal calculations later.

If the energy harvest was over unity, the discovery would shake the foundations of orthodox science.

* * *

Later in the lab, McKane felt the effects of being pummeled by a physical force made of light. The side of his face and arms ached from the split-second encounter with the Quantum Kinetic Repulsor, as if he had been stuck by Archimedes' giant hand after entering the forbidden harbor.

Unexpected things happen on quantum's event horizon, he reflected.

McKane realized that if he remained working in the unfamiliar harbor, in addition to clean and everlasting energy, he might discover other unimagined wonders on the alien continent. How many other explorers would follow him onto untouched territory, daring the risks and rewards of illumination. That was the outstanding question.

He sat down at the lab bench and began writing a restricted patent, his way of sending a message — and a map — back from the edge of the quantum frontier.

FIGURE 1: At the edge of the North American continent, a solitary portal points to a brighter horizon *as if* reminding humanity of the Ocean's limitless potential to transform end points into new life.

END

EPILOGUE

Nature and nature's God, in their infinite wisdom,
have evolved a species that is able to modify its own genes,
and that species happens to be ours.

Walter Isaacson
The Code Breaker

Note from a Secret Laboratory, Washington Coast, Fall 2025

It was against all the odds.

An automobile hobbyist using only his own hands, imagination, and a box of garage scraps discovered long-hidden truths about physics that could stop climate catastrophe in its tracks.

Suddenly, the amateur tinkerer had discovered *Cold Nuclear Fusion.*

The truth of it had made the young man's head spin.

It was a breakthrough akin to modern-day alchemy — the fantastical power to modify atomic elements with the turn of a simple electronic dial. The awe-inspiring technology offered our species the terrifying power to modify the Earth's environment and thus increase probabilities of human survivability in a threatened ecosphere.

Regrettably, the only objective problem for the inventor in propelling his remarkable machines into commercial space has been fierce opposition by experts in science and academia who had officially declared cold fusion a "pathological science" in 1989 because of the flawed measurements in the work of Pons and Fleischmann.

Thirty-five years later, countervailing peer reviewed research with Quantum Kinetics in three national laboratories had verified production of new nuclear products, the transmutation of atomic elements at room temperature, the release of high levels of excess energy, and cost-effective methods to overcome the massive force of the Coulomb Barrier. These premier, third-party studies had substantiated objective evidence against the classification of cold fusion as "pathological science."

Still, no one wants to admit they were wrong.

***

Quantum Kinetics' current aim is to grow the evidence of Cold Nuclear Fusion as a promising solution to the world's greatest problems. McKane Lee has put his personal life and future on hold to devote all his energy into proving the vast potential of his fledging fusion machines.

Despite the constraints inherent from a lack of institutional and outside corporate support, Lee and the Quantum Kinetics team press on in their private mission to travel beyond the quantum horizon.

ACKNOWLEDGMENTS

This is a true story about a young inventor who veered away from traditional scientific methods to pave a pathway for novel systems to advance human progress. The characters are drawn from real life events and encounters, although some of the people's names are obvious pseudonyms for reasons of confidentiality.

McKane would like to thank the long list of supporters, partners, friends, and family who played a role in his improbable journey as a next generation innovator. And Bulter for his constant companionship. Garry Vandekieft for his quiet wisdom. Mark Dreke for his ability to see the big picture. Anthony Russo for his friendship and fervent dedication to the truth. Randy Kent for his discerning judgment and ability to accurately assess forensic evidence. Michael Leach for seeing things deeply and quickly. Roger Emerson for his magnificent legal work. Randal Bird for his sixth sense of tuning into the ringing resonance of better angels.

McKane would also like to thank Matthew Fleagle whose literary experience and eagle eye helped with final editing of the manuscript.

Most of all, he is forever thankful to his father, Riley, and his mother, Victoria, for their unwavering support and guidance.

Authors' Note: To further highlight the ways in which the public can be included in the romance and intrigue of scientific discovery, we are already working on Episode III of *The Quantum Kinetic Chronicles*.

ABOUT QUANTUM KINETICS

Quantum Kinetics (QKC), world leader in nuclear fusion technology, is the first known company to successfully trigger and permanently sustain plasma fusion temperatures of 200 million degrees Celsius (indicated by 18 keV X-rays) and confirmed by independent analysis at three national laboratories. QKC has artificially decayed Uranium 234, 235, and 238 and transmuted Strontium-90, Cesium-137, and Molybdenum-99, all of which are cancer-causing nuclides found in waste from nuclear fission power plants.

These revolutionary achievements have been measured in peer-produced and peer-controlled studies, combining the experimental and analytical efforts of an impressive assemblage of physicists and engineers, all recognized as national authorities in their respective areas of scientific expertise.

Despite long being scoffed at by the scientific community as folly, cold fusion has returned to the scene with the creation of Quantum Kinetics' Electro-Physical Transmutation technology, which generates a cold fusion plasma that carries and controls high levels of energy without the costs of thermal dissipation systems and robust exotic materials that are inherent to the function of competing fusion technologies. The form of energy released during E-PT® is real, quantifiable, and measurable. In the past five years, Quantum Kinetics has rigorously confirmed and exhibited the repeatability of this fusion process. One of our peer-reviewed studies has been published in a <u>*Nature*</u> journal and was praised by the three reviewers as "*a much-needed invention around the world*".

Quantum Kinetics' accomplishments in bringing Safe Nuclear Energy Systems™ to the world will blaze the trail to abundant *zero-emission electricity. Additional developments are underway, to design and scale additional climate-remediating technologies, including* Used Nuclear Fuel (UNF) processing and the manufacture of high-value and scarce elements from low-cost inputs, both of which are necessary to the continued development of energy, transportation, and many other market sectors.

The economic implications of harnessing and controlling the power of the Sun in a safe environment are enormous. Quantum Kinetics Corporation's Star-in-a-Jar does just that. By re-imagining electricity generation with cold-fusion technology, QKC possesses the most promising technology for paving the way to a sustainable and environmentally friendly energy future that improves costs, profitability, reliability, and security for public infrastructure.

QKC's game-changing achievements have proven a highly promising pathway to realizing practical nuclear fusion and changing many of the world's industries.

Quantum Kinetics has an aggressive strategy to realize the world's first grid-scale commercial nuclear fusion power plants and nuclear waste remediation plants to seed multi-trillion-dollar industry.

Quantum Kinetics has no viable competitors in the present global marketplace for stable and sustained fusion energy, nor in the derivative industry of nuclear waste remediation through isotope editing. Competing commercial fusion companies have only achieved transitory bursts of fusion – often referred to as "ignitions" in homage of this problem – which range from a billionth of a second (at Lawrence Livermore) to the more impressive results obtained at the China Experimental Advanced Superconducting Tokamak (EAST), where a record-setting reaction was sustained, in 2021, for 101 seconds at ~120 million degrees Celsius (indicated by 10 keV X-rays).

Despite these advancements in these standardized and long-refined fusion technologies, Quantum Kinetics's results stand supreme, with 18keV x-ray emissions – indicating energy output equivalents in excess of 200 million degrees Celsius – sustained for longer than 24 hours. This was achieved without damage or noticeable weathering or wear in our reactor system. These results have been verified by third-party experts and showcase incomparable promise of a near future with grid-scale and affordable fusion energy.

QKC, with its Safe Nuclear Energy Systems™, Electro-Physical Transmutation (E-PT®), and Repulsors™ (Fast Fusion) technologies, stands to eventually re-write the way business is done in almost every industry. Transportation, mineral refining, water filtration, energy production, fusion mining, cancer treatment research, and more can quickly become exciting industry sectors where new applications of fusion technology are being developed. The potential applications of this technology are so wide-ranging that they truly are limited only by our imaginations.

QKC's unique know-how places it at the forefront of the S-curve for bringing fusion energy to the world, blazing the path forward for safe, clean, and globally deployed nuclear fusion technology.

The Genesis of Quantum Kinetics

McKane Lee's passion for science began with a deep fascination for quantum field theory, gravity, electrolysis, and hydrogen dynamics – subjects he explored intensely in long, self-directed hours. Inspired by rigorous science and science fiction, Lee devoted seven years working from his garage. He crafted and iterated on small experimental rigs built from scraps, wires, and salvaged electronics, relentlessly testing the mechanics of quantum and nuclear behavior under low-energy inputs.

The result of those efforts was the discovery of the Quantum Kinetic Well®, an exciting new technology that efficiently initiates and sustains fusion reactions. This technology produces nuclear reactions – at room temperature – without traditional hot-fusion's requirements for high-temperature plasma confinement and enormous power inputs to initiate and maintain transitory fusion reaction cascades.

A Vision Brought to Life: The Naming of the "Arc Reactor"

In a nod to his creative roots, McKane Lee named his prototype the "Arc Reactor", directly inspired by the fictional device from the Iron Man films. Yet it is not a simple aesthetic homage – the Arc Reactor is now patented, trademarked, and, most critically, a functional cold-fusion reactor system capable of sustaining nuclear fusion reactions even under ambient environmental conditions, and in water or air reaction matrix (this can also be extended to vacuum conditions). Operating at low voltage (often < 40V), this device produces soft X-rays (0.7–10 keV) and higher-energy emissions extending up to 30 keV, confirming low-energy nuclear activity through Bremsstrahlung radiation, among a litany of likely isotopic transmutation signatures. These emissions are key indicators of quantum-scale energetic behavior achieved without the fusion industry's conventional extreme temperatures and pressures or the dramatically expensive superconducting electromagnets and associated cryogenic infrastructures.

Breaking Barriers: From Garage to Global Aspiration

Lee's early breakthroughs did not gain immediate academic recognition. His research, initially overlooked due to its unconventional presentation and low-budget origins, eventually garnered attention through empirical rigor – X-ray spectra and isotopic modification of experimental samples treated under the watchful gaze of academic and industry experts.

These results confirmed that Lee's prototype could maintain plasma temperatures of 150–200 million °C for hours and longer – at less than 5 Watts of input – achieving a record-setting sustained nuclear fusion reaction.

Founding of Quantum Kinetics Corporation

Driven by the amazing potential of safe, low-energy nuclear fusion, Lee formalized his efforts into Quantum Kinetics Corporation, assuming the role of founder and CEO, along with co-founders Riley Lee, Micheal Leech, and Randy Kent. The company now holds multiple patents and trademarks in its growing IP portfolio. With Lee at the scientific helm, QKC has focused its developmental efforts on modular applications of: E-PT, Safe Nuclear Energy Systems, the Quantum Kinetic Fusor, and the Arc Reactor.

A Narrative of Perseverance and Imagination

McKane Lee embodies the archetype of a sci-fi-inspired inventor – part scientist, part dreamer. The books *The Quantum Kinetic Well* and *Journey Beyond the Quantum Horizon* articulate this narrative: a journey from obscure experiments to practical engineering, overcoming skepticism and technical constraints to create what are, in essence, functioning "star-building" devices, using Nature's own strategies for creating new elements in a gentle and continuous fashion.

Lee's message is clear: With a relentless spirit and attitude, along with outside-the-box thinking, even despite a shoestring budget, one individual can seed disruptive technology. "Anyone can change the world with a box of scraps," Lee often urges determination and creativity are the fuel of innovation.

QKC's Emergent Mission & Trajectory

Through its founding under Lee's leadership, QKC aims to democratize Safe Nuclear Energy Systems — bringing the benefits of scalable, modular, low-energy nuclear fusion systems to every industry. Lee stands at the nexus of this mission, melding imaginative inspiration, scientific rigor, and entrepreneurial intensity, setting the tone for QKC's ongoing journey from garage prototype to a global technological revolution.

Founders and Executive Board

Randy Kent, MS, P.E. (Founder)
Principle Engineer & Founder of Kent Engineering. Kent is a Registered Professional Engineer with over 41 years of experience in failure analysis and material engineering forensics. His expertise has been sought in developing the industry-standard ASM Handbook series, International Fatigue Congress Publications, with hundreds of depositions, trials, and arbitrations to his name.

In addition to his work at Kent Engineering, Randy is a professor of materials engineering for upper-level engineering students at the University of Washington, as well an advisor of curriculum development for Haward Technology Middle East. Randy holds a bachelor's degree and master's degree in metallurgical engineering, both from the University of Washington.

In his free time, Randy can be found with his wife, kids, and grandkids. He loves all adventures but especially enjoys fishing, hunting, and exploring on his boat.

Michael Leach, MS, P.E., CWI, CWEng
(Founder)
Director of Inspection & Engineering at Kent Engineering

Michael is a mechanical engineer with over 9 years of engineering and project management experience. Michael leads forensic consulting projects, facilitates site and laboratory investigations, and conducts finite element analysis (FEA). As a certified welding engineer (CWEng), he has managed and provided technical support on a wide variety of material failure cases.

Trained and mentored by industry leader and renowned welding engineer, Jay Dwight. Michael now leads Kent Engineering's welding testing, analysis, and engineering service group with the support of our in house materials and corrosion expert team.

Prior to consulting, Michael was a Design Engineer at a heavy machinery and agricultural machinery manufacturer in North Dakota overseeing new designs, product improvement, R&D, and client communication and relationships. Michael also played 3 years of Division 1 baseball at NDSU.

McKane Lee (Founder & CEO)
M.A., B.A. Global Cold Nuclear Fusion
Educator

Formerly a teacher and digital media professional with Electronic Arts, McKane Lee graduated with a Masters in Digital Media from Simon Fraser, Emily Carr, BCIT, and UBC. The Quantum Kinetics' self-taught, inventor, and missile-man leads with his imagination. Lee's original and divergent thinking outside the traditional physics box have led to the discovery of sustainable Cold Nuclear Fusion — joining science fiction and real-world physics with ARC REACTOR technology. Pablo Picasso was right about the power of unconventional thinking, "*Learn the rules like a professional, so you can break them like an artist.*"

Riley Lee (Founder & CFO)
Clinical Psychologist & Trial Attorney
B.A., M.S., J.D.

Before Riley Lee became a writer/biographer, he was an Attorney/CPA with over 33 years of experience in civil litigation, business, and tax law. As a trial lawyer, he won several top jury verdicts and settlements in Washington state. Prior to his legal career, Mr. Lee was a Clinical Psychologist for 20 years and also worked in human resource administration including as a county-wide Mental Health Manager. In recent years, Mr. Lee has transitioned from careers in psychology and law to become an author, utlilizing his diverse professional background, understanding of human psychology, and legal writing skills to delve into the dynamics of scientific discovery at the level of the individual inventor.

Randal Bird (President & COO)
Project Management Professional

Randal has over 40 years of experience in diverse business enterprise settings including aerospace, health insurance, life, property, and casualty insurance, and in the health provider markets.

Randal's main contributions to Quantum Kinetics Corporation are his skills in executive leadership, strategic development, program management, digital arts, and multi-media development and deployment. As a long-standing successful Project Management Professional, Randal's strong leadership comes from natural abilities, educational instillment, and nurtured professional practice.

As President and COO of QKC, Randal chairs the board meetings, ensuring forward direction in company development with oversight and accounting of scheduling and performance progress, budgeting, and partnering with the board in laying out the strategic vision.

Randal enjoys life with Ginny, his wife of 38 years, and their 4 children and 5 grandchildren. Randal is an avid daily exerciser and plays bass guitar. Randal has served over 17 years on the Board of Directors at his church, and over 10 years on the Board of Directors for Jumpah.org, a Cambodian humanitarian non-profit organization.

Bryton Rodgers (CPO)
B.S. Construction Manager
Industrial and Production Designer

Bryton has spent his career in land development, construction, and manufacturing, where he has built his close familiarity with materials sciences, along with a strong sense for coordinating the engineering, estimating, planning, and management of complex and technically challenging projects. Because of his diverse and active background, his skill set is uniquely tailored to understanding the practical needs of QKC's technological platform and orchestrating the skillful execution of the manufacturing and construction processes that will be required to bring cold fusion to market.

<u>Fundamentals of QKC Technology</u>

As another entry in the annals of the long-standing American tradition of revolutionary technologies being born in the mind and garage of the proverbial "little guy", Quantum Kinetics Corporation has recently emerged from obscurity with a technological innovation that will serve for the betterment of the condition and comfort of humanity as well as the health of our shared planet. Thanks to McKane Lee's ingenuity and enterprise, QKC was founded and has the patented and trademarked Quantum Kinetic Well (QKW). The varied and productive ends to which this technological artifice and its associates will be applied are well beyond the scope of this single document. However, early forays into testing and development of the technology have yielded strongly positive internal confirmation of these potential applications for QKC's technologies:

- Nuclear Fusion Reactors
- Power Generation
- Enrichment of Fissile Materials
- Accelerated Decay of Used Nuclear Fuels
- High Energy Hydrogen Generation and Storage
- Efficient Water Purification

All of these ends can be met with the use of any water as the dielectric matrix (or "fuel") for continuous, controlled, and safe modification of atomic nuclei within the modular and scalable fusion systems that QKC plans to deploy within 24 months after primary large-scale funding.

Quantum Kinetics has trademarked names for the following applications of these technologies, which are protected by patents and careful management of trade secrets:

- Quantum Kinetic Well (QKW)
- Quantum Kinetic Fusor
- Electro-Physical Transmutation (E-PT)
- Quantum Kinetic Repulsors
- Quantum Kinetic Fusion
- Quantum Kinetics Corporation
- Quantum Kinetics Corporation (Logo)

Below is list of qualitative research consultants that we have worked with to establish supporting evidence for our technological claims.

- Pacific Northwest National Laboratory – DOE
- University of Washington - MAF
- Kent Engineering (Data Curation Acquisition)
- NPJ Nature Clean Water Article
- <u>Continuing Research & Development</u>

QKC has engaged in thorough research and development efforts to verify the fact of the nuclear fusion reactions that can be initiated and sustained in the Quantum Kinetic Well, or through the processes of E-PT® or Safe Nuclear Energy Systems™.

Ongoing research efforts show promise for the development and refinement of these applications of QKC technologies:

- High-Density Power Generation
- High-COP Power Generation
- Stimulated beta-decay of toxic UNF
- de novo and Modified-Isotope Manufacturing
- Exotic Materials Manufacturing
- Waterway Cleansing and Separation of Fission waste

For further information visit: http://www.quantumkinetics.co
Youtube: https://www.youtube.com/@quantumkineticscorporation

www.ingramcontent.com/pod-product-compliance
Lightning Source LLC
LaVergne TN
LVRC081519210926
848486LV00017B/401/J

9798218788995